SpringerBriefs in Ethics

For further volumes:
http://www.springer.com/series/10184

Lisa Newton

The American Experience in Environmental Protection

Springer

Lisa Newton
Fairfield University
Shelburne, VT
USA

ISSN 2211-8101 ISSN 2211-811X (electronic)
ISBN 978-3-319-00049-7 ISBN 978-3-319-00050-3 (eBook)
DOI 10.1007/978-3-319-00050-3
Springer Heidelberg New York Dordrecht London

Library of Congress Control Number: 2013932094

© The Author(s) 2001, 2013
This work is subject to copyright. All rights are reserved by the Publisher, whether the whole or part of the material is concerned, specifically the rights of translation, reprinting, reuse of illustrations, recitation, broadcasting, reproduction on microfilms or in any other physical way, and transmission or information storage and retrieval, electronic adaptation, computer software, or by similar or dissimilar methodology now known or hereafter developed. Exempted from this legal reservation are brief excerpts in connection with reviews or scholarly analysis or material supplied specifically for the purpose of being entered and executed on a computer system, for exclusive use by the purchaser of the work. Duplication of this publication or parts thereof is permitted only under the provisions of the Copyright Law of the Publisher's location, in its current version, and permission for use must always be obtained from Springer. Permissions for use may be obtained through RightsLink at the Copyright Clearance Center. Violations are liable to prosecution under the respective Copyright Law.
The use of general descriptive names, registered names, trademarks, service marks, etc. in this publication does not imply, even in the absence of a specific statement, that such names are exempt from the relevant protective laws and regulations and therefore free for general use.
While the advice and information in this book are believed to be true and accurate at the date of publication, neither the authors nor the editors nor the publisher can accept any legal responsibility for any errors or omissions that may be made. The publisher makes no warranty, express or implied, with respect to the material contained herein.

Printed on acid-free paper

Springer is part of Springer Science+Business Media (www.springer.com)

Preface

Greetings all, and welcome to the growing world of Environmental Ethics. What follows in this introductory text is an account of the growth of environmental awareness and environmental conscience, in the course of the twentieth century (brought up-to-date in the twenty-first as appropriate), primarily in the United States. I have aimed to keep it short and readable, for maximum usefulness. The major limitation occasioned by this aim is its parochialism: all the cases and literature selected are of the United States (the case of the lethal explosion in Bhopal, India, was a US case, and held implications only for US policy); all the weights and measures are in pounds and miles. The greening of American thought, as seen in public attitudes toward the natural environment, may generalize to the world at large, but the peculiarly American interaction among the peculiarly American institutions in the course of this history suggests that indigenous social and cultural expectations may play a large role in any nation's orientation to the natural environment that surrounds their history.

Several audiences guided the composition of this work. First, it is directed to undergraduate and graduate students, whom I have had the great pleasure of teaching (and learning with) for over 40 years, and who I have found to be earnest and delightful but woefully ignorant of the background, concepts, and defining conflicts in the field of environmental studies. We often felt the need for a short treatment of the key cases, initiatives, and literature that created this field. Whether the students will be doing bench science on the diseases of plants or fieldwork on the effects of climate change in the southern Appalachian Mountains, I knew they would do better with a feel for the history that led them there.

I was especially concerned for our business students, who were headed off to the corporate world with training primarily in economics, finance, management, marketing, and demonstrations that reliance on the free market will solve every problem. They tend to learn that environmentalists are their enemies. They are not; American corporate leaders have become some of the strongest advocates for the preservation of the health of the land. For this reason, I have included several of the cases that have shaped the relationship between the major actors in the movement to preserve the environment and the corporations that found themselves, not by their own choice or expectation, ranged in opposition to them. This text distinguishes itself from the bulk of the environmental literature by ensuring that the perspective of the corporation is fairly represented.

Another audience, following on that point, may be the corporate officers who find themselves (not necessarily by their own choice or expectation!) responsible for developing or implementing environmental policy within the corporation. Possibly they have time to educate themselves in the sciences that comprise ecology and the history of corporate confrontations, but likely they do not; this book should tell them all they need to know. For that matter, it should provide enough fact and insight to help any citizen sort out the political/environmental choices before us. After all, they have a right to know.

The need for brevity and rapid absorption in the various audiences has led, inevitably, to a concise organization that must constitute oversimplification for purposes other than my own. It has also led to decisions to omit entirely consideration of social ecology, ecofeminism, bioregionalism, and other primarily political aspects of the subject. It has long been my contention that the various topics of "environmental justice" have nothing to do with the environment and everything to do with larger legal and political fields of justice; they await another work.

Anyone, scientist or otherwise, who has tried to follow the concerns for the biosphere, warnings of climate change, loss of water, and numerous others, knows that scientific facts—number of parts per million (ppm), number of degrees of global warming to be expected under various scenarios, number of people whose health may be at risk from environmental conditions, population figures—form the bulk of most arguments. I have not always tried to include the latest figures on these problems; by the time the book reaches the shelf in the store, they would be out-of-date. It will always be the student's job to update the numbers. But I have tried to state the problems, put them in perspective, and encourage the reader to find out more on his or her own.

The book is now yours, to read, mark, and reflect upon; I hope you have as much joy reading it as I have had in writing it.

Contents

1 **Environmental Background** 1
   Keynote for the Field: Aldo Leopold. 2
   Case 1: Hetch Hetchy Valley 3
   Case 2: Prospectors and Neighbors 4
   Case 3: Rachel Carson, DDT and the New Threats to Our Health 5
   Reflection: How Had We Come to This Point? 6
   Case 4: Malthus Revisited: Kenneth Boulding's Spaceship, Garrett
   Hardin's Commons and Paul Ehrlich's Population Bomb 8
   Case 5: Norman Borlaug and the Green Revolution 12
   Case 6: Earth Day and Environmental Legislation 13
   Case 7: Barry Commoner's Laws of Ecology 14
   Case 8: Love Canal and Toxic Wastes 15
   Case 9: Bhopal and Industrial Safety 19
   Reflections on the Democratic Process 22
   Case 10: The Wreck of the Exxon Valdez 24
   The Sacrifices for Oil and the New Focus on the Ecosystem 24
   Further Reflection: The Reentry of the Ecocentric 28

2 **Business Hears the Call** 29
   Case 11: The Business Case for Environmental Protection:
   3M & NYC 30
   Case 12: Green Marketing and the American Consumer 30
   Sustainability and the Triple Bottom Line 32
   Case 13: The New Industrial Revolution and the Automobile 33
   Case 14: The McMansion 35
   Case 15: Biomimicry and Beyond 36
   Reflection: The Philosophers 37
   Biodiversity 39
   Case 16: Biophilia? Circling Toward Home 41
   Reflection: Deep Ecology and the Justification
   of Monkey Wrenching 41

3 Coming to Value Nature 43
  Man as a Plain Citizen of the World: Return to Aldo Leopold 43
  Ethics Focused on the Natural Environment 50
  Eight Ways of Seeing 52
  The Case for Environmentalism 59

Introduction

There is little consensus on the nature and power of environmental ethics. One significant historical fact underlies all discussions: where only a few centuries ago Nature—the non-human part of our world—was regarded as nothing but an obstacle to our purposes and sometimes a means to our ends, an alien, an enemy to be conquered; Nature is now seen as the basis for all human as well as non-human life, valuable and worthy of preservation. The conditions for this transformation are generally known: with the explosion of the human population and its expansion to all habitable parts of the Earth, the boundaries of what is still wild are plainly seen; with the overuse of the fisheries and forests, the limits of the natural resources upon which we have depended are now in sight; and a series of natural collapses and catastrophes, threatening to the health of humans and others, have brought home to us the need to adopt a protective attitude toward the natural ecosystems.

Implicit in this transformation is a profound change in perception, in the way non-human Nature is seen and comprehended. All other ethics are anthropocentric, oriented to the good of human beings. Part of the subject matter of environmental ethics starts there: Nature is good, and must be protected, for the sake of human safety (from poisons, toxic spills, and explosions), human health (vs. pollution of air, water, and soil), and human enjoyment (in the preservation of the scenic wonders of the wilderness). Against the anthropocentric subject matter, there stands the ecocentric or biocentric part of the environmental ethic: Nature is good for its own sake, and should be protected, all of it, just for that reason. No reasons need be advanced to ground such an obligation to protect, any more than reasons have to be advanced to support the value of an individual human life, and the duty not to kill. Between these two clusters of values, there is a third, calling on the spiritual aspect of the natural world: Nature is beautiful, and requires our attention for that reason alone, and for what it does for our souls when we are in contact with it.

Chapter 1
Environmental Background

Abstract The growth of "environmentalism" in the United States is traced from the earliest efforts to preserve the American wilderness through the politicization of environmental crises, real and perceived, to Earth Day and the national commitment to protect the natural environment and human health. The initial opposition between private enterprise and an orientation to the natural environment is briefly sketched, and the emerging obligation to shepherd products from "cradle to grave".

Keywords John Muir • Aldo Leopold • Rachel Carson • Garrett Hardin • Norman Borlaug • Barry Commoner • Love Canal • Bhopal • *Exxon Valdez*

What values, dilemmas, and conundrums form the field of environmental Ethics? What does the earth, its water and air, mean to us? What are the rights and duties of humans *vis a vis* the natural world from which they came? The cases and materials of an environmental ethics course tend to center around three sets of values: human health (physical and emotional), human economic welfare (profit and loss), and the flourishing of the land itself. The first two values are *anthropocentric,* measuring value, benefit and harm, in any situation, against human well-being of some sort. Incidentally, they tend to conflict: generally, human health is threatened by a polluted environment, so in the interests of health we must minimize pollution; but pollution (in the classic cases) tends to arise from profitable business enterprise, which improves the economic welfare of the people. In the interests of jobs and the economy, then, while never praising pollution, we ought to be very cautious in our efforts to abolish pollution for fear of hurting business. The third value is *ecocentric,* measuring the worth of any policy or practice according to its effect on the ecosystems of the earth. The original, and clearest, ecocentric ethic is the "land ethic," first and most eloquently advanced by Aldo Leopold in the mid-twentieth century: **"A thing is right when it tends to preserve the integrity, stability, and beauty of the biotic community. It is wrong when it tends otherwise".**[1] All anthropocentric ethics can be subsumed, if indirectly, into general ethics (through a series of hypothetical imperatives); but at the heart of

---

[1] Aldo Leopold, *A Sand County Almanac and Sketches Here and There*, New York, Oxford University Press, 1949, 224–225.

environmental ethics is *ecocentricity:* the basis of the ethic, and the center of moral value, is the earth, the land itself, the created physical world of which humans are a very small part. Ultimately the question of environmental ethics is *why* the earth is valuable all by itself, and there are no easy answers.

Keynote for the Field: Aldo Leopold

Aldo Leopold (1886–1948) was a graduate of the Yale School of Forestry (1909) who worked for the U.S. Forest Service for nearly twenty years, primarily in predator control (hunting wolves, usually). In his *Notes from Here and There* he tells the story of his change of heart, from happy hunter to reflective environmentalist. It's worth repeating:

> We were eating lunch on a high rimrock, at the foot of which a turbulent river elbowed its way. We saw what we thought was a doe fording the torrent, her breast awash in white water. When she climbed the bank toward us and shook out her tail, we realized our error: it was a wolf. A half-dozen others, evidently grown pups, sprang from the willows and all joined in a welcoming mêlée of wagging tails and playful maulings. What was literally a pile of wolves writhed and tumbled at the center of an open flat at the foot of our rimrock.
>
> In those days we had never heard of passing up a chance to kill a wolf. In a second we were pumping lead into the pack, but with more excitement than accuracy: how to aim a steep downhill shot is always confusing. When our rifles were empty, the old wolf was down and a pup was dragging a leg into impassable slide-rocks.
>
> We reached the old wolf in time to watch a fierce green fire dying in her eyes. I realized then, and have known ever since, that there was something new to me in those eyes—something known only to her and to the mountain. I was young then, and full of trigger-itch; I thought that because fewer wolves meant more deer, that no wolves would mean hunters' paradise. But after seeing the green fire die, I sensed that neither the wolf nor the mountain agreed with such a view.[2]

Leopold goes on to describe a landscape browsed to death by a surfeit of deer, until, every shred of vegetation gone, their bleached bones join the denuded landscape. In order to preserve Nature, he insists, even the part most useful to human beings, we must learn to "think like a mountain," with the whole ecosystem in view. His classic *A Sand County Almanac* and other essays proceed from this point, culminating in the statement of the "land ethic," above.[3] (Note that the Land Ethic answers the question, "why is the earth valuable?" It makes the earth the center of value, and measures human activity by its effects on the earth.) It would be wonderful if American environmental consciousness had taken off from there. But Leopold, by the end of his life a quiet professor at the University of Wisconsin, was little noted nor long remembered by the American public at that time. (His name recognition has improved since then.) The Land Ethic may have

---

[2] Aldo Leopold, *A Sand County Almanac and Sketches Here and There*, New York: Oxford University Press, 1949 pp. 129–130.

[3] *Ibid* pp. 224–225.

been the earliest comprehensive statement of environmental ethics in the United States, and it may be where we end. But we have a long history of crises to go through before we get there.

Let us begin with some of the more familiar examples. All these cases are real (reality overtook my ability to construct scary scenarios years ago). Notice, as we go through these cases, that while the burden of the case (and why it made it into the literature) is anthropocentric, focusing on the ability of people to stay healthy, make a living or enjoy themselves, even the simplest cases call upon ecocentric dispositions, as reinforcement for the anthropocentric conclusions.

Case 1: Hetch Hetchy Valley

Environmental consciousness had begun, for the United States, in the ninetieth century, when the gentlemen of the hunting class discovered that their hunting and hiking grounds were being eaten up by the industrial revolution's insatiable demand for wood. President Theodore Roosevelt's progenitors founded the first National Park, Yellowstone, in 1872, not to go shooting, but in large part for the love of Nature and the desire to preserve it for enjoyment, theirs and that of all the American people. In the establishment of such Western destination points, they had the support of the railroads, looking for tourist dollars in an expanding market for rail travel. Hikers and businessmen were joined by the visionaries of a new preservationist movement, led by an eccentric adventurer, John Muir. Muir had been blinded in an industrial accident as a boy, moved west to the mountains and fell in love with them. He vowed that should he ever regain his sight, he would dedicate his life and work to preserving those mountains, and spearheaded a move to keep them forever pristine and protected. He did regain his sight, and began his campaign, which brought Yosemite National Park into being in 1890, but he didn't stop there. He formed the Sierra Club in 1892, to promote exploration and consequent protection of the wild areas, to promote the responsible use of resources, and above all to educate the citizenry on the value of protecting environmental quality and on the ways to accomplish that protection. Meanwhile Roosevelt's friend Gifford Pinchot, founder of the Yale School of Forestry, noticed that unlimited logging was threatening our resources for the future. The Forest Management Act of 1897 undertook to identify and protect forest reserves, primarily for their economic uses, to save the wood, water, and mineral resources for later use. These reserves became National Forests in 1907. Save for this reservation, that there were some forests that we were saving to cut later, the creation of forest reserves was due in some part, always, to love for the beauty of nature, with no further reason necessary.

To review this early trend in environmentalism, then, we started with a love for the land, the land as it was, for its beauty, for its wealth as a resource, and for our own enjoyment and spiritual refreshment. Note from the beginning, the fault line in the conservationist movement. According to Gifford Pinchot, Teddy Roosevelt's

Forester, the woods were an economic resource for the United States, needed by her people, whose future the President and he served. According to John Muir, no human purpose could justify destroying the natural beauty of the woods themselves. The two understandings of the move to preserve the forests were bound to lead to conflict at some point, and they did monumentally, in Yosemite Park's Hetch Hetchy Valley. The problem was simple: San Francisco and its surrounding region needed more water; the waters of the Tuolumne River, swelled by snowmelt from the surrounding mountains, would provide the region with more than enough pure water; and the river ran through a conveniently stable and narrow valley at Hetch Hetchy. So Pinchot proposed to dam the river and send the water to San Francisco. From 1901 to 1913 Muir and the Sierra Club conducted furious letter-writing and lecturing campaigns against the dam, arguing both that the wilderness had a right, preceding human rights, to persevere in its original form (Muir), and that the beautiful valley should be preserved for the joy of future humans (most of the other letter-writers). The effort did not succeed. In 1913 Woodrow Wilson signed into law the Raker Act, authorizing the damming of the river. It took 20 years to build, but in 1934 the Tuolumne Reservoir started delivering water to San Francisco. That water system now serves 2.4 million citizens, plus businesses and public facilities of all kinds. If the current movement to "Restore Hetch Hetchy" prevails, where will water come from?[4] This conflict is real, and is in no way resolved.

Case 2: Prospectors and Neighbors

The environmentalist move to restore the valleys occasionally worked. Not every set of dams supplied entire cities. Some had been built, on the smaller rivers, just to run local mills, now out of date and closed. The environmental damage of these dams, unforeseen or disregarded at the time they were built, was significant: the salmon and steelhead runs were interrupted, sometimes ended, there was no way to replenish the downstream environment with upstream nutrients, and recreational uses of the river, fishing, kayaking and rafting, were impossible. One such river was the Rogue River, flowing through Oregon to the Pacific Ocean. It took the NGO Water Watch 23 years of campaigning and lobbying (and lawsuits) to bring down four dams that blocked the river from Eagle Point to Grants Pass on the river, but they won—the last dam, Gold Ray, was demolished August 11, 2010.[5] Wonderful! The river now belongs to nature and to those who would innocently enjoy it!

---

[4] The information came from the websites of the Sierra Club, of Restore Hetch Hetchy, and the Bay Area Water Supply and Conservation Agency, on July 1, 2007.

[5] Felicity Barringer, "Where Dams Once Stood, Prospectors Spur Anger," *The New York Times*, September 4, 2010, A10.

Case 2: Prospectors and Neighbors 5

Not so fast. The next sound heard on the river was not the babbling of water over stones, or the scream of the fish-hawk, but the deafening roar of gas-powered gravel dredges, as prospectors seeking gold clustered below where the dams had been, dredging up mounds of gravel from the bottom, sifting through the piles for flakes of gold, then returning all the gravel to the bottom in huge mounds, only to start again a few feet away. After all this work to restore the river, does the environment now belong to the "New 49'ers" [sic] as they call themselves? Immediate efforts were made to determine how much environmental damage they were doing, with (as is usual in these cases) mixed results. To be sure, the dredges destroyed the bottom, where the fish spawn, and the dredges can suck up eggs and small fry; but the dredging happens in summer, when the spawning is done. It destroys insect nests on the bottom; but these return in weeks. Meanwhile, fish get to eat the disrupted insects. All the dredgers have permits from the State of Oregon Department of Environmental Quality; Beth Moore, general permit coordinator for the state, "said 1,205 dredging permits had been issued this year, up from 934 in 2009".[6] That's revenue for Oregon, with no obvious environmental damage, except for the assault on the ears of the neighbors, and possibly other damage to them—the neighbors claim, for instance, that the silt is fouling their irrigation pumps. At first it was assumed that the prospectors would only be active on weekends—but this year, they've been at it seven days a week. No real economic profit from the mining is alleged—but the hope of gold can keep the activity going through a long run of dry holes. The only claim on the part of the prospectors is that they have that hope, that they want to prospect for gold in this way, that they are American citizens, that they have permits, and therefore that they are free to ignore the neighbors. They see themselves as "citizens whose rights are under siege".

Who are the adversaries here? So far, the ecosystem does not figure in; permits can be drawn to restrict the seasons of prospecting so the salmon are not hurt. This particular battle is between groups of citizens whose interests in interacting with the environment just happen to be incompatible (like those of the hikers who want to enjoy the natural ambience of the mountain park, and the motorcyclists who want to use the trails for adventurous and high-decibel riding). Who should have precedence? This is a matter for local politics—suppose it were the neighbors who were doing the dredging?—and any solution will have to be negotiated repeatedly.

Case 3: Rachel Carson, DDT and the New Threats to Our Health

The period immediately following World War II in the United States was defined by the triumph of Science. The changes that affected the American land in this new industrialized time were enormous. Plastics were discovered. Mighty

---

[6] *Ibid,* p. A10.

machines and household appliances, consuming electrical or fossil fuel energy rather than human effort, transformed the home and the workplace. We believed in miracles, benign miracles. For the first time in history, we could contemplate the possibility of eradicating disease and feeding the world. These hopes were spawned by two of the banner scientific discoveries of the mid-century: antibiotics for the infectious diseases that used to kill our children, and organochlorines to rid the world of the insect pests that had carried disease and decimated our food crops since the beginning of human existence. Foremost among these chemicals was the pesticide dichloro-diphenyl-trichloroethane (DDT), apparently harmless to human beings (the son of its developer recalls traveling the country dramatically eating the stuff to prove it wouldn't hurt anyone), but absolutely deadly to insects—we sprayed the malarial swamps and the lairs of the tsetse fly, and we also sprayed our farms, to create the rich tapestry of efficient industrialized agriculture, which could produce more food than we had ever dreamed of.

It was the DDT that launched the current wave of environmentalism—the anti-industrial revolution, I suppose we could call it. The era was born in 1962, with the publication of *Silent Spring,* Rachel Carson's masterpiece.[7] Rachel Carson was a biologist by profession, a naturalist by persuasion, with many excellent portrayals of nature in and near the sea to her credit. But the theme of this book was not that we ought to preserve nature for its beauty, but that our imposition of industrial methods and profit motives on nature, in the form of the widespread use of the pesticide DDT, was death-dealing in its effects on non-target species. Humans, for instance, at the time she wrote, carried large burdens of DDT in their own bodies. We had thought it was harmless, but it was not. In the body, pesticides mimic chemicals in the endocrine system, and may have serious negative effects on reproduction. As she wrote, the reproductive failures of the magnificent raptors, our prized eagles and falcons, which fed on the birds and fish contaminated by the insecticide, had already been documented.[8] That our industrial agricultural chemicals might be dangerous to higher species had been suggested for decades, but Carson had the proof. Slowly and reluctantly, the Congress of the United States brought itself to intervene in the Free Market system to save the birds, and that was the beginning of environmental legislation in the country.

Reflection: How Had We Come to This Point?

What created the environmental crisis? In 1967, Lynn White jr published his groundbreaking essay, "The Historical Roots of Our Ecologic Crisis," in *Science* magazine, one of the most prestigious journals on the planet. His argument was that the major culprit of destruction was the theology of human domination of the

---

[7] Rachel Carson, *Silent Spring,* Boston: Houghton Mifflin, 1962.

[8] Carson, pp. 118–122.

earth, originating in the Book of Genesis chapter 1, where God assigns to Adam the power and duty to name, i.e., dominate, all the other animals. All of creation, on White's reading of the tradition, was subordinate to human desires; "God planned all of this explicitly for man's benefit and rule: no item in the physical creation had any purpose other than to serve man's purpose".[9] This understanding became part of Christian assumptions; my Christian university has cut in stone (literally, around the top of the columns of the porch of one of the dormitories) the quotation from St. Ignatius, "God made man [humans] for Himself, and everything else in creation for man".

To the Judeo-Christian tradition, White joined the religion of Technology. According to what we now know as the Technological Imperative, humankind has always exhibited an unchecked enthusiasm for each new device, opportunity, machine, ever larger and more complex, to increase productivity and consumption of the world's resources. From the invention of the wheel to the crafting of the atomic bomb, each "advance" in human technology had rapidly become irreversible, laying the groundwork for the next.

The Technological Imperative is not a surprising emergence in the human experience. Long before Capitalism and the Free Market came on the scene, competition for resources was a fact of life. There has never been a time when the resources available to a community were not "scarce," from the perspective of the members of the community, who had to divide them up somehow. (We need not retreat to Darwinian times, and the competition for dominance and for mates; in any discussion of the twentieth century environmental movement, we may assume that the players have social structures, like elections and proposals of marriage, which channel such conflicts). Technological advance—whether in the form of a better plow, or tractor, or turbine, or high-speed railroad, or jet aircraft, or computer (or iPhone), or method of creating financial products (like derivatives) without the U.S. Securities and Exchange Commission noticing—has always advanced the economic fortunes of its possessor, and so written itself irreversibly into history. We do not expect its early disappearance.

White's critics rallied first on the theological issue. There are no plans for an environmental movement in the Bible, but there are many places where the natural world is celebrated and loved. Further, White's interpretation of Genesis is open to question. It is not clear that "naming" means "dominating": it might just mean organizing into an intellectually available framework for human enterprises. And in any case, the *second* chapter of Genesis gives a much clearer picture of the desired relationship between humankind and Nature: first God created the Garden, then "God placed man in the Garden *to tend and to keep it*". The relationship is not one of domination, but of stewardship—it is the human's job to protect Nature, the garden, *and to give an account to God, as to any owner, for the conduct of his stewardship at the end of days.* Worth noting is Psalm 24: The earth is the Lord's,

---

[9] Lynn White Jr., "The Historical Roots of our Ecologic Crises," *Science*, 155, 1967, p. 1205.

and the fullness thereof, the world and they that dwell therein. It does not belong to humans. (Note that the passage renders any human "right" to private property problematic.)

The Technological Imperative is similarly open to question. Its major role is, as above, as an aid to competition, which is ordinarily assumed to be necessary to survival. But it doesn't have to be. If social arrangements can manage distribution of scarce resources according to some plan accepted by the whole community, competition becomes unnecessary (in fact counterproductive), and the Technological Imperative can be scrapped. At this point the community can decide exactly what level of technology they want to accept, what kind of work they want to do and what kind of relationship they wish to maintain with the larger community. (There is an impressive literature on the strategies for forming communities that practice voluntary simplicity, even voluntary poverty.[10] All of them consciously reject the Technological Imperative.)

On the Christian imperative to dominate nature, then, White is simply wrong, or at least incomplete. On the Technological Imperative, he is probably right, historically, but there are alternatives, to some of which we can return at the end. But White's thesis was unusually important in the history of environmentalism: for the first time, scholars were invited to examine their own intellectual and cultural history for the sources of the problems they addressed. According to Baird Callicott, most of the philosophical discussions of the natural environment in the 1970s were dedicated to debating Lynn White's thesis.

Case 4: Malthus Revisited: Kenneth Boulding's Spaceship, Garrett Hardin's Commons and Paul Ehrlich's Population Bomb

In 1966, not that many years after Rachel Carson's pioneering work, economist Kenneth Boulding published "The Economics of the Coming Spaceship Earth," a powerful Malthusian argument that the resources of the earth, overused by an increasing population, were severely limited and decreasing. For the entirety of human existence to this point, he argued, humankind had enjoyed the illusion (and for the first several million years of human existence, the fact) that the material and energy resources of the earth were limitless in relation to human needs. Now suddenly we were looking at a closed and limited system, finite in resources and possibilities, putting firm boundaries on human enterprise. It was a bit like the closing of the frontier. Even if we avoided poisoning ourselves, we have to worry

---

[10] See, for instance, Linda Breen Pierce, *Choosing Simplicity,* Carmel, CA 2000; Mark A. Burch, *Stepping Lightly,* Gabriola Island, BC: 2000. Also good is the history and literature of The Catholic Worker, founded by Dorothy Day.

about how much we could use; if you have to live on a spaceship, every mouthful of food, indeed every breath, must be accounted for.

Powerful as Boulding's work was, Boulding was not the one who succeeded in raising the alarm about the expanding population in a world of diminishing resources. That honor belongs to Paul Ehrlich, who was a scholarly entomologist at Stanford University when (possibly inspired by work with insect populations) he added up the numbers and came to the conclusion that there were too many people on the earth. His calculations and his 1968 book *The Population Bomb*, follow those of Thomas Robert Malthus, whose *Essay on the Principle of Population as It Affects the Future Improvement of Society* (1798) cast a pall over the optimistic legacy of Adam Smith; unless halted by predation, war, or pestilence, population of any species continues to increase geometrically while the food supply even under the best of circumstances can only increase arithmetically. Therefore every species will outbreed its food supply until starvation brings its increase to a halt. Humanity, specifically, would double in number until it ran into one of several end-stage scenarios—nuclear wars, forced sterilizations, the ultimate survival of cockroaches alone.[11]

Not only humans would suffer in the end-time of over-population. As all the resources of the earth were exhausted, used up by increasingly thoughtless industrial expansion, the soil itself, drained in the futile efforts to grow enough food for everyone, poisoned by the overdoses of chemical fertilizers and pesticides poured on it for the same purpose, the last generations of humans would contemplate a dead planet. We were headed for disaster, and something had to be done. He suggested sterilants in the water supply, financial incentives to limit children, and above all a guaranteed right of abortion. *The Population Bomb* (1968), followed two years later by a more comprehensive (and better documented) textbook to the same effect, *Population, Resources, Environment: Issues in Human Ecology*, attracted a wide following, started the Zero Population Growth movement, and sparked a relentless campaign of opposition from the Catholic Church (whose teachings on abortion and contraception Ehrlich attacked directly), and eventually from a variety of minority groups and advocates from the developing world, who saw in his recommendations a campaign of pre-emptive genocide. In his final definitive work, *Ecoscience: Population, Resources, Environment,* joined by his wife Anne and his colleague John Holdren, Ehrlich continued the argument, drawing in data from more fields and significantly dividing the field of debate into two positions, the "cornucopians" (who believed firmly in the ability of technology to solve problems long before the apocalypse) and the "Neo-Malthusians," who did not, putting himself firmly in the latter camp.

Also in 1968, a banner year for environmentalism, Garrett Hardin brought out his famous essay, also in *Science,* "The Tragedy of the Commons". His argument was simple: think of the world and its resources as the "commons" in the middle of a sixteenth century village, where the flocks of the villagers were grazed in an area

---

[11] See John Hersey, "My Petition for More Space".

protected from local predators, primarily wolves. The commons was necessarily limited in space, and could accommodate only so many sheep—in this assumption, Hardin followed Boulding's spaceship. Every villager had an equal right to graze sheep on the commons, and to harvest his sheep whenever he wanted to, to monetize their meat and wool. Clearly it was to each villager's interest to graze as many sheep on the commons as possible. Of course, if everyone added to his flock without limit, the commons would soon be unproductive, and there would be no grazing for anyone. That commons is a perfect model, he argued, for the seas, superbly rich with fish and other delicious food when the first colonists arrived in New England, now largely gone—even as he wrote, the cod that had been so thick on Georges Bank and the Grand Banks only years earlier, were headed toward near-extinction. It seems that they will not come back. What to do about the tragedy? Hardin asserted that the Commons problem in sixteenth century England had been solved by dividing up the commons into private plots, since individually owned productive property will be well conserved by its owner in his own interest, and the problem of overuse will no longer occur.

In 1974, Hardin elaborated on the implications of his Commons analysis in "Living on a Lifeboat." Like Boulding's Spaceship, Hardin's lifeboat was severely limited in resources, and required strict accounting of all consumption. Unlike the spaceship, Hardin's lifeboat had people swimming around in the water, soon to die, begging to be allowed to come aboard. But if the folks in the lifeboat allowed others to climb in, the lifeboat would surely run out of resources very soon, and probably sink before that. Therefore no one should be allowed on; the present passengers in the lifeboat should pick up the oars and row away, leaving the others in the water to die. The lifeboat modeled the world: its passengers were the citizens of fortunate lands with abundant food, good water, comfortable homes, good health, and access to all the riches of the earth; the swimmers were the rest of the world, dying in a sea of starvation, disease, envy and despair. We might want to help them, but unfortunately we just can't.

Hardin also ventured into a critique of classic economic philosophy, oddly interpreting the utilitarian key value of "the greatest good for the greatest number," as advanced by Jeremy Bentham and J.S. Mill, as recommending an increase in the number of humans without limit. That was not what was meant, as Hardin's critics immediately pointed out; the phrase in question is completed with "of the population in question," and had nothing to do with an increase in population. Hardin was more severely taken to task for the influential "commons" idea. We understand a commons; common ownership precedes private property by several millennia, and its outlines and rules are clear. Any commons—land, tools, animals, hunting rights, whatever—is collectively administered by the organization that claims it by right, and the rules for its care and use are worked out and administered by that organization. If a shepherd decided that it was to his advantage to add some more sheep to the commons without approval from the organization, and did so, he might find himself contemplating some dead sheep in the morning. The ocean is not a commons, although portions of it can become one; notoriously, the coastal lobster fishery off the state of Maine is administered by the local lobstermen, and a stranger who arrives in the waters and drops traps without contacting the locals first may find his buoys adrift in the morning. (This bit of local administration may have changed with the advent of the GPS; I am told most lobstermen now do not use buoys, but locate their traps with their GPS and drop grappling hooks to retrieve them. I don't know what they do to trespassers now). The ocean otherwise is an "open access system," not a "commons," available to all without limit. There are no restrictions on the size of the boats that may be used for fishing, no restrictions on the processing of the fish, no matter how wasteful, and no restrictions on the amount that may be taken—the more, the better. What is needed is surely not an attempt to "privatize" the ocean (maybe by dividing it up into nine billion equal plots?), but an effort to establish an enforceable international system of rules, quotas and monitors arranged to make sure that the stocks of fish are conserved while fishermen take enough for their needs. (Incidentally, there is no guarantee that the owner of private property will take good care of it). We know this quota system can work, as the Coast Guard successfully monitors the annual salmon runs along the Alaska coast. It is not easy or fun: the size of the fishing boats is set by law, and enforced; if a boat strays over the set line while fishing, the Coast Guard gunboat has the right to order that boat out of the water for the season, and a year's income is lost. Further, all the boats and the Coast Guard are U.S.; how such a system would work with boats of every nation and a blue-capped UN enforcement navy (or Coast Guards from all over the world?) is very unclear. But the fact remains, that the fisheries are all but destroyed, and that something has to be done if this crucial food supply is to be preserved.

In a concession to the cornucopians Ehrlich agreed that if technological ways could be found to increase the food supply, possibly his predictions might be defeated. One such cornucopian, Julian Simon of the University of Maryland, publicly disputed all of Ehrlich's claims. He insisted that human beings were better off now than ever before, that historically, the more enterprise, the higher the technology, and the more humans in the population, the greater the increase in human welfare. Humans were our greatest resource. Ehrlich had argued that as more people put more pressure on the resources of the earth (the raw materials used in industry), the resources would become scarcer and accordingly more expensive. Simon challenged him to a bet: Ehrlich predicted that the prices of certain materials would rise, Simon predicted that they would fall due to technological advance: let us set a date to compare the prices and the loser will pay the winner the difference between the predicted and actual prices. Simon won, and Ehrlich had to pay him $576.07.[12] Boomsters defeat Doomsters. Their victory was underlined by an unexpected expansion of agriculture, the "Green Revolution."

---

[12] De Steiguer, J.E. *Age of Environmentalism*, New York: McGraw-Hill, 1997, p. 90, citing Charles McCoy, "When the Boomster Slams the Doomster, Bet on a New Wager," *The Wall Street Journal*, 225 (108) Monday, June 5th, 1995, pp. A1, A9.

Case 5: Norman Borlaug and the Green Revolution[13]

Norman Ernest Borlaug (1914–2009) earned a doctorate in plant pathology and genetics from the University of Minnesota in 1942, and supported by the Rockefeller Foundation, went on to develop high-yielding disease-resistant strains of semi-dwarf wheat in Mexico. Combining these new varieties with contemporary agricultural practices (including fertilizers and pest control) he made Mexico a net exporter of wheat by 1963—in that year, the wheat harvest was six times what it was in 1944. Transferring his work to famine-prone Asia, he was able to help India and Pakistan double their wheat production by 1970. In all the developing nations that adopted the new strains, production increased by a factor of three between 1970 and 2004, making the hungriest nations in the world self-sufficient in cereal grains. Again with the support of the Rockefeller and Ford Foundations, he turned his attention to rice, and within the five years between 1965 and 1970 brought the Asian acreage in high-yield wheat and rice from 200 acres to 40 million acres. These advances, in countries racked by hunger, have been called the "Green Revolution," and were thought to be the answer to hunger in the world.

Borlaug argued, throughout his life, that the Green Revolution was the best protection for the natural environment. There is broad agreement that saving tropical rainforests around the globe is crucial for the future of the environment and all forms of life in it; since the major reason for cutting down the rainforests is the production of more agricultural land, he argued, increasing the productivity of the land we already have is the best way to protect the forests. (Critics immediately pointed out that there are many reasons to cut down the trees in the rainforests, profit from lumber trade in the rare woods being the major one, and there is no way better agriculture could affect that enterprise). Other environmental critics directed their fire against the Green Revolution's encouragement of monocultures, hence of large-scale industrial farming, the imposition of capitalist agriculture on traditional societies, and the imposition of single-strain wheat or rice on the enormous diversity of varieties of traditional farming. Borlaug dismissed these criticisms as elitist: the complaints of the very well fed about the means, taken by the chronically hungry, to feed themselves. Hunger hurts, he pointed out, and should any of his critics have to live for even a year in the hunger he had witnessed for fifty years, they would be crying out for tractors and artificial fertilizer. Ultimately, his critics may be right; the Green Revolution agriculture makes very heavy demands on water and fertilizer, and when these are lacking, the crops fail. Irrigation carried on too long can also waterlog the soil and leave salt on the surface; evidence suggests that this process ended the abundant agriculture in the ancient Fertile Crescent.

But he had made his point. The Malthusians were wrong; agriculture could raise the food supply to the point where everyone could be fed (should the rich, who overeat, take it into their heads to put into place an efficient distribution

---

[13] For an interview of Borlaug discussing the Green Revolution, see http://www.actionbioscience.org/biotech/borlaug.html.

system and share their food with the poor). However the human race might fare on the earth, it was not going to starve to death, as a whole, any time soon.

Case 6: Earth Day and Environmental Legislation

Within a remarkably short time after *Silent Spring*, environmental consciousness on many other matters was raised all over the country. Suddenly citizens noticed that you had to turn on your headlights at noon in Pittsburgh, that there were days in Los Angeles when you could not breathe the air, and that the Cuyahoga River in Cleveland was on fire. When day turns to night and the rivers burn, absent some signal from Gabriel's horn, we knew that things were not going right with this world. What's happening to us? And what can we do about it? All the writers above reacted to the crisis, so recently noticed. The new concern came to an unprecedented culmination in the celebration of the first Earth Day in 1970, proposed by Wisconsin Senator Gaylord Nelson. The Baby Boomers, the healthiest and best educated generation in history, came on the scene 20 million strong, determined and in force. Galvanized into action, Congress passed landmark legislation to protect the environment—the Clean Air Act, the Clean Water legislation (the Water Quality Improvement Act, and significant amendments to the Water Pollution and Control Act), acts requiring recycling (the Resource Recovery Act and the Resource Conservation and Recovery Act), the Toxic Substances Control Act, the Endangered Species Act, the Marine Mammal Protection Act, many others—and created the Environmental Protection Agency and the offices needed to make it work.[14] The natural environment in the United States came under highly effective protection, the best in the world, and in the process the pattern was set—industry may have no responsibility of its own to work to preserve the environment, but where harm to higher forms of life are suspected, confrontation of industry by corporation-disliking citizens will elicit command and control from the government.

The attacks on Rachel Carson launched in the period immediately following the publication of *Silent Spring* by the chemicals industry foreshadowed the unhappy future of this moment; they included personal denigration (with open sneers at the ability of women to do science at all), scorn for her scientific credentials and methods, and ultimately, a total disregard for the truth. Ezra Taft Benson, Secretary of Agriculture under Eisenhower, is credited with the note wondering "why a spinster with no children was so concerned with genetics?" He was not alone in deciding that she was "probably a Communist".[15] The pesticide industry trade group, the National Agricultural Chemicals Association, spent well over a quarter of a million dollars (about $1.4 million in today's dollars) in campaigns

---

[14] The legislative history is too long for this essay; that brief list was excerpted from Gaylord Nelson, "Earth Day'70: What It Meant," in the EPA Journal, 1980. Found on the website http://www.epa.gov/history.

[15] Linda Lear, *Rachel Carson: Witness for Nature*, New York: Henry Holt, 1997, p. 429.

against Carson's conclusions, her work in general, and her, as an emotional female alarmist. They saw, more accurately than most, that if her work was generally accepted, they would not only lose pesticide sales immediately, but also public trust over the long term, and would face massive regulation. In all this they were right.[16] That's what Earth Day was all about.

Case 7: Barry Commoner's Laws of Ecology[17]

Barry Commoner, a Professor of Biology at Washington University in St. Louis, wrote his best-known work, *The Closing Circle: Nature, Man and Technology*, in 1971, at least in part in reaction to the uncertain trumpets of environmental passion broadcast on Earth Day, which he thought could use some clarity. The Circle referred to in the title represented for him the normal operations of Nature, processing matter and energy through an endless productive cycle, providing food for all and neatly disposing of all waste. Humans, he felt, had broken the circle, putting new demands on resources, living out a greedy agenda of acquisition, transformation, production and consumption that steadily diminished the resources of the earth available to all other creatures. We had forgotten the basic laws of ecology, which Commoner proceeded to enumerate:

1. **Everything is connected to everything else**. Everything in nature is dependent on everything else, and any disturbance in one part of it will have effects somewhere else. What this means, essentially, is that we can never do just one thing. Just spraying the garden with insecticide to kill the mites that are eating the vegetables, we also kill the predator insects that would have kept down their numbers, possibly the birds that eat the insects (Rachel Carson's case), and an unknown number and kind of the microorganisms that inhabit the soil and keep it alive. The consequences down the road are beyond our power to predict.
2. **Everything Must Go Somewhere.** Or as the popular press put it, "there's no such thing as away," into which you can throw the garbage. The most troublesome form of garbage was the toxic chemical mess left over from the earlier dumps.
3. **Nature Knows Best.** Virtually all human-initiated changes in the ecosystem turn out to be harmful in the long run.
4. **There is No Such Thing as a Free Lunch.** Every gain must come from somewhere, somewhere that will suffer as a result. Prominent among the illustrations for this law is the Green Revolution, above, where enormous gains in the food supply depleted the soil and required petrochemical-based fertilizer.

---

[16] *Loc. Cit.* In general, for the industry reaction to Silent Spring see Lear, Chapter 18, "Rumblings of an Avalanche."

[17] Read Commoner's book, *The Closing Circle: Nature, Man, and Technology*. New York: Knopf, 1971.

Case 8: Love Canal and Toxic Wastes[18]

The 16 acre Love Canal site is located in Niagara Falls, New York, destination of honeymooners and one of the unluckiest cities in the United States. In 1892 William T. Love began the development of an industrial site along the canal that connected the Niagara River to Lake Ontario, realized he couldn't make any money on it, and abandoned it, the canal half dug. The Hooker Chemical Company took over the isolated, abandoned canal in 1947, having received permission to use it as a waste disposal site in 1942. The company had complied with what few requirements were necessary for waste disposal at that time, and was confident that the site would not leak. The canal bottom consisted of a soil primarily clay, the preferred liner for waste sites, because it is virtually impermeable to water. (Because some chemicals will diffuse through it—a three foot clay barrier will leak mobile chemicals in five years—most modern sites use a mixture of clay and synthetic materials). By 1952 the company had dumped 21,800 tons of chemical wastes into the site. Hooker was not the only source of chemicals in that dumpsite, by the way; several Federal agencies, especially the army, arranged with Hooker to dump residues of wartime production in the same spot. Meanwhile, residential building had begun nearby, as the city of Niagara Falls expanded outward. In 1953, the canal could accept no more waste, so the company covered it, again, with a cap of solid clay.

Soon afterwards, an expanding population required more schools in the area, so the company sold the site (reluctantly, and under threat of condemnation) to the Niagara Falls Board of Education for $1.00. Apparently at that time the School Board wanted that site only for a playground, and that was fine with Hooker. They made sure to insert into the deed (dated April 28, 1953), a strong disclaimer regarding any injury to come from those wastes:

> Prior to the delivery of this instrument of conveyance, the grantee herein has been advised by the grantor that the premises above described have been filled, in whole or in part, to the present grade level thereof with waste products resulting from the manufacturing of chemicals by the grantor at its plant in the City of Niagara Falls, New York, and the grantee assumes all risk and liability incident to the use thereof. It is therefore understood and agreed that, <u>as a part of the consideration for this conveyance and as a condition thereof,</u> no claim, suit, action or demand of any nature whatsoever shall ever be made by the grantee, its successors or assigns, against the grantor, its successors or assigns, for injury to a person or persons, including death resulting therefrom, or loss of or damage to property caused by, in connection with or by reason of the presence of said industrial wastes. It is further agreed as a condition hereof that each subsequent conveyance of the aforesaid lands shall be made subject to the foregoing provisions and conditions.[19]

---

[18] For a history of the Love Canal tragedy, see http://www.epa.gov/history/topics/lovecanal/01.htm.

[19] Deed of Love Canal Property Transfer, Niagara Falls, New York, April 28, 1953. Emphasis added.

So matters stood for four years; then the School Board decided to build on part of that site and sell the rest. Hooker showed up at the hearings; company officials were insistent that hazardous wastes were under ground, and issued dire warnings about what might happen if that clay cap were pierced. The minutes of the School Board's meeting on November 7, 1957, indicate that A. W. Chambers of Hooker was present at the meeting, specifically warning about the dangers of those buried chemicals should the site be disturbed. He conceded that the company had no further control over the use of the property, but urged strongly that none of that land be sold or used for building houses or other structures.

When the School Board eventually decided to build an elementary school on the site, then, the company had informed the Board that chemicals were buried there; they had not, however, told them what the chemicals were, or to what extent they were toxic; nor did they tell them what quantity of chemicals were buried there.[20] It is not clear that company officials knew any of those items of information. A later review of the situation revealed quite an assemblage:

> What lay beneath the surface was 43.6 million pounds of 82 different chemical substances: oil, solvents and other manufacturing residues. The mixture included benzene, a chemical known to cause leukemia and anemia; chloroform, a carcinogen that affects the nervous, respiratory and gastrointestinal systems; lindane, which causes convulsions and extra production of white blood cells; trichloroethylene, a carcinogen that also attacks the nervous system, the genes and the liver... The list of chemicals buried in the Love Canal seems endless, and the accompanying list of their acute and chronic effects on human beings reads like an encyclopedia of medical illness and abnormality.[21]

Trouble could be expected. An elementary school was built, and house lots were sold. As early as 1958 children were burned from playing in the dump, probably from the pesticide lindane, some 5,000 tons of which Hooker had buried there, and which surfaced in a cake-like form.[22] Then in 1977, chemicals started to appear in the basements of nearby houses after heavy rains, leaching out of their grave like some apparition of ghosts on Hallowe'en. Michael Brown, a reporter for the *Niagara Gazette* wrote of resident complaints of dizziness, respiratory problems, a chemical stench, breast cancer, and pets losing their fur.[23] In 1978, national publicity prompted both state and federal action.

First of all, what did they find there? 200 chemicals were identified on the grounds around the school built over Chambers' protests: among them benzene was prominent, and is credited for initiating government action. New York

---

[20] Conversation with Eugene Martin-Less, Esq., Environmental Bureau of the Attorney General of New York State, Albany, New York. Senior Attorney, New York State versus Occidental Chemical Company (suit pending), asking $250 million punitive damages relative to Love Canal dumpsite.

[21] Nader, Ralph, Ronald Brownstein, and John Richard, eds., *Who's Poisoning America: Corporate Polluters and Their Victims in the Chemical Age.* San Francisco: Sierra Club Books, 1981.

[22] Whitney, Gary. "Hooker Chemical and Plastics," *Case Studies in Business Ethics,* ed. Thomas Donaldson, Englewood Cliffs, N. J.: Prentice Hall, 1984.

[23] Michael H. Brown, "A Toxic Ghost Town Harbinger of America's Toxic Waste Crisis," *The Atlantic,* vol. 263, #1, July 1989.

Case 8: Love Canal and Toxic Wastes

State began studies of the site, supplementing the newspaper anecdotes prevalent to that time. A panic atmosphere, however, is not ideal for scientific studies. Before any real study, Robert Whalan, New York State Commissioner of Health, declared an emergency and moved to evacuate about 20 families. On August 9, 1978, Governor Hugh Carey visited the site and declared that all 236 families living along the streets affected would be permanently relocated, at state expense. (At that time a new clay cap was placed over the old Canal area). As residents of the houses just outside those purchased and evacuated by the state began to find more and more ailments among them, Dr. Beverly Paigen, a biologist with Roswell Park Memorial Institute in Buffalo, came out with her own study of a few families and their ailments. Her evidence was largely anecdotal, but triggered enough interest to bring the Environmental Protection Agency into the case.

Early in 1980, the EPA commissioned a study by Dr. Dante Picciano on chromosome damage in the area, which claimed to find elevated damage. Hard on the heels of this study came another, by Dr. Steven Barron, on nerve damage among inhabitants of the Love Canal area; he also claimed to find elevated levels. The wake of these two reports, coming at a time of tension as they did, featured a full-scale riot, including the taking of two EPA officials hostage by the local Homeowners' Association. Within days, President Jimmy Carter declared a State of Emergency at Love Canal, announcing the relocation of another 2500 residents—first temporarily, at a cost of $3–$5 million, then permanently, at a cost of $30 million.

Given the nation's lack of experience, the Love Canal evacuation was a model of effectiveness. New York State capped the canal with clay and installed a drainage system that pumped any leaking material to a new treatment plant, started demolishing the most contaminated houses, and started buying up the others that had been abandoned. In 1982 and 1983 the houses, the school, the parking lot, and the original playground were all demolished. By 1990, everything that is west of the street that backed up to the canal, for one-quarter mile, had been buried and fenced off, and the State owned 789 single family homes. The total clean-up costs at that point were estimated at $250 million.

On the other hand, there may have been no danger. None of these studies—not Beverly Paigen's anecdotal study of epilepsy, mental illness and reproductive difficulties (including miscarriages) in a few families, not Dante Picciano's studies, which lacked controls, nor Steven Barron's study, which was admitted to be a pilot study but which was leaked when the others became public—has withstood scientific scrutiny. How can we tell whether chemicals in the environment have caused disease? Perhaps a word about scientific testing is in order.

The determination of toxicity is difficult and expensive, and unless we start experimenting on humans, the results will always be controversial. The most common technique is animal testing, although there are techniques that use bacteria and other organisms or tissues. The smaller the population tested, and the less time taken to significant results, the less expensive the test, so these tests typically are conducted on animals—frequently white mice bred for genetic sameness—which are exposed to a much larger quantity of the chemical, over a much shorter period of time, than any human population would ever experience. The lethal dose to all, the lethal dose to half (LD_{50}) and the dose that causes no effect are recorded and those figures extrapolated to humans. On the basis of such testing, the scientists who conducted the research were able to say that many of the substances found at the Love Canal site—including benzene, dioxin, toluene, lindane, Polychlorinated biphenyls (PCBs), chloroform, trichlorethylene, trichlorobenzene, and heavy metals—are certainly capable of causing harms. Thirteen are known carcinogens. With the exception of heavy metals, most of the chemicals found there would be liquid at room temperature and soluble in water, increasing the chances of migration away from the site. Additionally, chlorinated hydrocarbons are denser than water and would sink, migrating towards ground water.

It is very hard to prove cause and effect in these cases—that exposure to a given chemical at a given time causes a specific symptom. In order to determine that, it would be necessary to know the quantity of the chemical to which each sufferer was exposed and the period of time of exposure, for each individual within the exposed population. Additionally, testing should be repeated in a number of exposed individuals, other causative factors have to be ruled out, and the cause and effect should make physiological sense. There were 200 chemicals at Love Canal and an exposed population that occupied some 800 homes. Separating each chemical and symptom from others appears to be a monumental task; and given the number of chemicals, synergistic and antagonistic reactions cannot be ruled out. There has never been a comprehensive study done on the health effects purported to have resulted from chemical exposure at Love Canal. And of the less-than-comprehensive studies that have been done, the ones that tried to connect the chemicals to the symptoms (those mentioned above) have, as above, not been accepted by the scientific community. Many of the early accounts, the accounts that drew the attention of the media to Love Canal, were informal surveys done by the residents themselves. The state and federal government studies mentioned above were among the few governmental studies done. As above, Paigen's results were anecdotal and impossible to interpret; Picciano's, on chromosomal damage, were uncontrolled. A group of 17 scientists from the Centers for Disease Control, Brookhaven National Laboratory and Oak Ridge National Laboratory, attempted a follow-up study on chromosome damage in 1983, and found that if anything, the chromosomes of Love Canal residents were healthier than the norm.[24] As for the miscarriages, those results were taken up specifically by Dr. Nicholas Vianna of the New York State Department of Health:

> Efforts to establish a correlation between adverse pregnancy outcomes and evidence of chemical exposure have proven negative. Comprehensive studies of three households with unusually adverse reproductive histories did not produce evidence of unusual risk of chemical exposure.... We have not yet been able to correlate the geographic distribution of adverse pregnancy outcomes with chemical evidence of exposure. At present, there is no direct evidence of a cause-effect relationship with chemicals from the canal.[25]

---

[24] "CDC Finds No Excess Illness at Love Canal," *Science* vol. 220, June 17, 1983.

[25] Vianna, Nicholas. Report to the New York State Department of Health. Reported in *Science*, June 19, 1981, p. 19.

Meanwhile anecdotal evidence continued to accumulate, including stories of seizures, learning problems, eye and skin irritations, incontinence, abdominal pains, lung cancer, non-Hodgkins lymphoma and leukemia. In children, birth defects, low birth weight, and hyperactivity were noted. One child born of parents in the area was deaf, had a cleft palate, deformed ears, a hole in the heart, and impaired learning abilities. But anecdotal evidence, as above, carries little weight in scientific circles. Even if it could be shown that the incidence of such ailments is statistically higher in the Love Canal area than elsewhere, and it has not been, no connection to exposure to chemicals could be shown. But the residents, as may be imagined, dismiss all denials of that type as politically motivated (like the Tobacco Institute's disclaimers on the link between smoking and cancer), and continue to insist that the New York State Department of Health (DOH) underestimated the health effects of the event. The DOH scientists, for their part, believe that the health effects were *over* estimated by unqualified independent investigators, and that the second evacuation in 1980 was unnecessary. There seems to be general agreement that the psychological toll on the residents has been immense. Any slight physical symptom becomes a cause for concern; as one (former) resident said, "it's like AIDS".

If the scientific and health data are uncertain, the legislative effects are not. No one disputes that the publicity given to the symptoms reported by the Love Canal residents and to the eventual abandonment of the area was the driving force behind Congress' enactment of the Comprehensive Environmental Response, Compensation and Liability Act (CERCLA or "Superfund"), designed to assess liability for hazardous waste sites, and to clean them up. By this act the EPA is empowered to sue the owner, or the dumper, for the clean-up costs; if the site is significant, the responsibility for payment is usually settled in court. (The other piece of legislation that controls hazardous waste is the Resource Conservation and Recovery Act (RCRA, 1976, 1984) which requires dumpers to obtain permits and describe how the material is to be treated. It also requires "cradle to grave" reporting of waste, from origin to final disposal. This requirement is generally acknowledged to be unenforceable, given the estimated 750,000 hazardous waste producers and 15,000 hazardous waste carriers).

Uncertainty surrounded the science, but the foundations of U.S. environmental legislation, the first stones of which were laid on Earth Day 1970, were now in place.

Case 9: Bhopal and Industrial Safety[26]

On December 2, 1984, a pesticide plant in Bhopal, India, built by Union Carbide Corporation (UCC) and run at that time by Union Carbide India Limited (UCIL), sustained a monstrous explosion. Forty tons of methyl isocyanate (MIC) gas had

---

[26] See also http://www.bhopal.com/ for more information.

blown up, releasing into the air surrounding the plant a lethal mixture of MIC, hydrogen cyanide, monomethyl amine, and carbon monoxide, among other chemicals. The figures are still in dispute and will probably remain so, but an estimate in March, 2000, put the dead at 3000 and the injured at 200,000.

The company whose gas exploded into that lethal cloud was largely Indian-owned and completely Indian operated. It had been founded (as a branch of the American corporation) almost 50 years ago to provide pesticides for India's agricultural "green revolution"; the plant at Bhopal dated from 1969. There was nothing exotic or extraordinarily dangerous in the operation of its plants; the most common kind of pesticide produced in them is carbaryl, an ester of carbamic acid, a reliable and relatively safe product, marketed in the United States under the brand name SEVIN. The chemicals employed in the process of making the pesticide *are* dangerous. Phosgene, the deadly gas briefly used in World War I on the battlefield (and also in the gas chambers of the Third Reich) is a precursor of SEVIN. The Union Carbide India Limited (UCIL) process for its manufacture uses phosgene ($COCl_2$) and a methyl (CH_3) amine (NH_2) to produce the intermediate compound methylcarbamoyl chloride ($CH_3NHCOCl$). The latter compound breaks down with heat into MIC and hydrochloric acid (HCl). Methyl Isocyanate (CH_3NCO) is a variation of the cyanide group $(NCN)^{-2}$) of which the highly poisonous hydrogen cyanide (HCN) is probably the most famous. MIC is extremely unstable and dangerous, and as such is not ordinarily studied in a laboratory situation. Its boiling point is 39 °C. (102.4 °F.) Lighter than water in liquid form, but heavier than air in gaseous form, it hugs the ground when released. Its breakdown products include carbon dioxide and stable amines (organic compounds of carbon, hydrogen and nitrogen), but the process releases a vast quantity of heat (exothermic). It reacts violently with water (producing breakdown products and high temperatures) whether it is the water that entered the MIC storage tanks or the water in human tissue. Therefore it is an extremely dangerous human poison, and there is no known antidote. Occupational Safety and Health (OSHA) regulations allow human exposure at 0.02 parts per million (ppm) over an 8 h period, irritation is felt at 2 ppm, and becomes unbearable at 21 ppm. 5 ppm will kill 50 % of an experimental rat population (LD_{50}). Of course no one measured the concentration of the escaped gas at Bhopal, but as 50,000 pounds of it escaped, the heart of the cloud must have greatly exceeded those limits. There had never been an incident releasing large amounts of this gas; the managers at the plant had no idea that it was dangerous.

It made a lot of sense to put pesticide plants in India instead of manufacturing the SEVIN in the United States and exporting it: transportation costs (and dangers) were eliminated, and labor costs were a good deal lower in India, making the whole operation safer and more profitable as far as Union Carbide was concerned. It also provided tax revenues and very good jobs in a chronically depressed economy, in consideration for which the Indian governments sought, welcomed, and catered to those American companies that were willing to locate plants in their large and needy country. The land on which the plant in question was built was given to Union Carbide by the Indian government for an annual rent of $40 per

Case 9: Bhopal and Industrial Safety

acre, as part of the plan to bring industry into the area. (Bhopal is the capital city of Madhya Pradesh, the largest and one of the poorest states in the nation). In practice, the divided ownership and consequent division of responsibility for the safety of the plant—the Americans responsible for the design, the Indians responsible for implementation—fostered an attitude of complacency and unconcern for the details of the safety arrangement, and of mutual suspicion for decision-making authority. Those attitudes predict the inexcusable inattention to safety lapses before the explosion and the tragic chain of events that followed the explosion: recriminations, litigation, continuing political hyperbole, threats of further litigation, no relief at all to the actual sufferers, and no success in restoring the environment.

For over a year after the explosion the government of India would not allow industry investigators to examine the scene of the accident and analyze the residues in the wrecked tank. When they were allowed to visit the site, the investigators found, among other junk in the tank, chloroform ($CHCl_3$). That was a contaminant; it should have been removed by distillation earlier, but could not be due to a higher than normal temperature in the still (the refrigeration unit was broken). The initial heat of the reaction had come from the simple interaction of water and MIC, an intensely exothermic reaction. But it was the chloroform that provided the chlorine ions that attacked the steel lining of the tank, which in turn released the iron ions that acted as a catalyst for what the chemists call "trimerization," (three molecules of MIC reacting with each other to form a more complex molecule), a reaction that is even more exothermic. MIC, usually held at 20 °C. (ideally at 4.5 °C) finally reached 120 °C. By now of course the MIC had boiled (vaporized), the pressure blew the tank, releasing a cloud that covered 40 Km^2. None of the safety devices had worked, and the emptying of the huge tank—and the resulting devastation—was inevitable.

Every environmental event becomes a political event before the sun goes down. Indian authorities immediately concluded that the United States was responsible, especially the U.S. corporation Union Carbide, and launched into a campaign of blaming. The U.S. reaction was very different. When the news of the explosion reached the U.S., the Union Carbide employees took up a collection for the victims' families, the best engineers were assigned to get to India in the next two weeks to track the cause of the disaster, and Warren Anderson, Union Carbide's CEO, got on the next plane to India with authorization to spend one million dollars immediately for the relief of the victims. But in the heated political atmosphere, Anderson was arrested as he got off the plane and jailed. He was not allowed to speak to anyone or offer compensation to the victims; then he was summarily sent home. The employees of UCIL, all Indian, were immediately dispatched to other work, making them hard to trace, and any evidence suggesting misconduct (for example, the evidence that the responsible officers had been taking a tea-break together, contrary to regulations) was covered up. The cause of the explosion was alleged to be some backwash from a cleaning operation, brought about somehow by U.S. negligence.

The investigating engineers quickly found confirmation for the causal explanation they had deduced from the core samples. An instrument supervisor from the plant, otherwise not involved in the explosion, had surveyed the area of the tank on the morning following the explosion, and found that a pressure gauge had been unscrewed from the tank and was missing. That would explain how water got into the tank. Further, a hose, normally used for cleaning, was still attached to the faucet not far from the tank, and water was still running out of it. That would explain the source of the water. Apparently further investigation uncovered the name of the employee who had performed this senseless act of sabotage. He had recently been, or was about to be, demoted, and he was angry. He surely had no intention of causing that kind of explosion. His family probably lived nearby. But he knew water would ruin the batch, and that's what he intended to do. Union Carbide investigators quietly turned their information over to the local authorities. In the light of what was occurring around them, they cannot have been surprised that those authorities paid little attention to it.

Reflections on the Democratic Process

The hostility that attended the Bhopal disaster was different in degree, but not in kind, from the rest of industry-environment interactions in the 1980s. The long dance of regulation had begun, and characterized the movement called "green" for the next decades. The script is familiar: Industry carries on its (probably environmentally harmful) activities for the sake of profit and the public be damned, so went the script; Greenpeace (or other NGO or citizens' group) would discover the activities, prove them harmful to human beings, and bring them to the attention of the public; strident debates among the parties follow in train, and amid the Congressional hearings, lobbying, and moves on the floor of the Senate, the legislature will come to some regulatory conclusion. Through this period of the campaign to save the natural environment, the posture of opposition, protestor versus industry, became so well known as to become defining; environmentalism was only recognizable as an attack on some industry or other, and any such attack became part of the "green" movement.[27] Private enterprise always damages the environment, in this drama, and government, the bigger the better, is its only protector.

---

[27] In the middle of a discussion of population problems and the natural environment, G. Tyler Miller, Jr. includes a sidebar ("spotlight") on the 1970s story of infants in the developing world suffering from malnutrition because of inappropriate use of infant feeding products, contaminated with the local water, or too expensive for the family to purchase in sufficient quantity. Nestlé S.A. was accused of inappropriate promotion of the products. The infant formula case never at any point had anything to do with the natural environment; the INFACT protest had only to do with the health of babies in the developing world. Nothing substantially connected the two discussions, save that an NGO was opposing an industrial giant. Miller, *Living in the Environment*, 11th edition, Pacific Grove, CA: Brooks/Cole, 2000. p. 272.

This dance continues to this day, of course. Even as a whole new age of environmental consciousness has begun, this confrontation continues: there is no way to escape the responsibility of national governments to protect the safety and health of its people. Industry does not always win, and Rachel Carson—ultimately, posthumously—won the pesticide battle, and set the precedent, that we will not tolerate assaults, intentional or otherwise, on the health and safety of people and animals. But the environment does not always win: in July, 2007, for instance, the Environmental Protection Agency, so brave in the 1970s, agreed not to enforce a broader range of wetland protection directives "after intense lobbying from property owners, mine owners and developers".[28] In the same year, the EPA refused a routine waiver that would have allowed California and seven other states to hold automobile emissions to a higher standard than that mandated by the federal government—a waiver that had been fiercely opposed by the automobile manufacturers. These incidents point up an ongoing weakness in the old dispensation: Government action at several levels is still the only way we can deal with problems of large areas and the long run; but government follows the election returns, and it is easier for an anti-regulation administration to undo the protections put in place by its predecessor, than it is for a pro-regulation administration to repair the damage done by the anti-regulation people that came before.

Before we move to the next phase of the human relation to the environment, it might be worthwhile to underscore that last point. We have a tendency, we political animals of the United States of America, to think that democracy will ultimately reach the right answer—that as the political pendulum swings too far to the right (in tax cuts for the rich and sharp restrictions of social programs), in the next election it will return to the left (in higher taxes or fees, a higher minimum wage and greatly increased support for health, education and welfare). Those who were pushed aside will be back in the center, those who did well for themselves will have to tighten the belts, and it will all come out even in the long run. But that won't do for the environment. When the deregulation permits the cutting of 2000 year-old redwoods, the toppling of the mountaintops for the coal mine, and the scouring of the sea-bottom for the last of the cod or the blue fin tuna, the succeeding tree-hugging liberals will not be able to swing any pendulums at all—what's gone is gone for good. Nor will "compromise," our great political philosophical heritage, do us any good at all when it comes to the natural environment. Shall we cut the trees for jobs and profit now, or preserve them for posterity? "Compromise" would suggest cutting half of them. But next year, the same loggers want jobs—last year's trees don't set this year's table—and the investors want this year's profit. "Compromise" will lose another half of the trees. This halving can go on until there is only one tree left, and as we know, one tree does not make a forest ecosystem.

---

[28] John M. Broder, "After Lobbying, Wetland Rules Are Narrowed," *The New York Times*, July 6, 2007.

Case 10: The Wreck of the Exxon Valdez

The Sacrifices for Oil and the New Focus on the Ecosystem

At 9:30 PM of March 23, 1989, the Exxon Valdez, a 987-foot oil tanker owned by the Exxon Shipping Company (subsidiary of Exxon U.S.A., which is part of Exxon Corporation), left the dock at the port of Valdez with a cargo of 1.26 million barrels of North Slope oil, brought in from Prudhoe Bay to the terminal at Valdez through the Alaskan Pipeline, and headed out through Valdez Harbor, for Prince William Sound, the Gulf of Alaska and onward to the open ocean.[29] At 11:25 PM, Captain Joseph Hazelwood called the Coast Guard to tell them that he was leaving the outbound shipping lane and steering a course South (180 degrees) into the inbound lanes (empty at that time), to avoid ice floes broken from the Columbia Glacier to the North. The channel as marked and as normally traveled presents no difficulties for navigation. But Hazelwood had planned a course that would skirt the southern edge of that 10-mile opening in order to avoid the ice, and the third mate left in charge of the bridge did not start to turn back into the channel in time. The ship hit a rock, then fifteen minutes later grounded on Bligh Reef. The reef punched eight holes in her hull, spilling a quarter of a million barrels of oil (about 11 million gallons) into the clear waters of the Sound.

Who was hurt by the accident? There was no loss of (human) life in the crash, thank goodness. But economically, there was a world of human pain in store. The fish were the most expensive victims. The fishing industry earns $100 million annually in Prince William Sound (out of $2 billion for all of Alaska). Herring and salmon are the mainstays, but the industry also relies on crab, shrimp, Pacific cod, Alaska pollock, rockfishes, halibut, flounder and sharks. In 1988, Prince William Sound salmon fishermen earned $70 million from a harvest of 14.9 million salmon, 15 % of the state harvest. King crabs (48,422 pounds in 1988), shrimp (178,000 pounds) and herring roe are equally valuable: herring roe sells for up to $80/pound in Japan, and the industry makes $13 million annually. The herring need clean kelp to lay their spawn; after the spill, the kelp was sure to be covered with oil. Most of the towns, and most of the independent businessmen on the Sound, make their living one way or another on the fishing industry, which employs 6000 people all by itself. It is a growth industry; the average American consumption of seafood has nearly doubled in the last 20 years. (Such losses are typical, and the extent of your employment in the fossil fuels industry may determine your ultimate evaluation of them. For instance, in Philip Revzin's reassuring article, "Years Temper Damage of Worst Oil Spill [the sinking of the Amoco Cadiz, off the coast of France]" we are told that "For a year after the gooey oil washed up... business [fishing, oyster-growing, tourism] all but stopped. For five to eight years the aquatic food chain was disrupted, costing crab fishers three

---

[29] For a description of the wreck, see http://www.eoearth.org/article/Exxon_Valdez_oil_spill.

generations of their most prized catches". And oil remained, in pockets. "But generally, 'it's finished, all is back to normal,'" says Lucien Laubler, chief scientific advisor to a French oceanic research institute. Getting "back to normal" may become more problematic as the spills continue). Cleanup operations in Prince William Sound were delayed by lack of preparation, as the companies scrambled to find booms and skimmers. Then a winter storm tore through the area on the fourth day after the spill and spread the oil from 100–500 miles^2. As the oil continued to advance, it became clear that the precious salmon hatcheries were in danger. Nothing was being done to protect them. The fishermen were afraid that even a few gallons of that oil, sweeping through the rearing pens, could wipe out an entire generation of pink salmon. So they took their own boats and went to work, dipping oil and laying booms to protect their hatcheries. They were only moderately successful.

The Exxon Valdez spill was a major blow to the incomes of the Alaskans of the area. But that aspect is not what most troubled even the immediate commentators. For the first time, the greatest part of the accounts focused on the tragic assault on the ecosystem itself of Prince William Sound. Every account of the spill includes reference to the "pristine" beauty of the Sound before the accident: the wealth of birds, fish, wildlife, kelp, living things of all sorts; the complex web of interdependence that makes a threat to any of those species a threat to them all. Prince William Sound had never been fouled, polluted, cut over, industrialized, or settled by environmentally intrusive human groups. This incident was seen as all the more harmful because there was no precedent, not of damage, and not of recovery.

Prince William Sound is roughly 70 miles long and 30 miles wide, with many bays, inlets, and islands to break up its shoreline. There are approximately 2,500 square miles of open water, 1,800 miles of mainland shoreline, 1,200 shoreline miles on islands and rocks. The depth ranges from 2850 feet at its deepest, averaging 480 feet in the shipping channel—to virtually nothing over Bligh Reef at low tide. It is bounded by rocky peninsulas and towering mountains, most prominently the glacier-covered Chugach Mountains to the north. The wildlife is diverse and abundant. The Sound is normally home to several species of marine mammals, including sea lions, whales, seals, porpoises, and sea otters; many species of land mammals, hundreds of species of birds, including swans, cormorants, millions of shorebirds, and over a thousand bald eagles. All of these were at risk. It is also a major crossroads for migration; millions of migratory birds stop over in late April and early May, including one-fifth of the world's trumpeter swans. One bay in the Sound is home to the largest concentrations of orcas, or killer whales, in the world; the sea otters make up perhaps a quarter of the total U.S. sea-otter population; the marshes and estuaries near Cordova on the eastern side of the sound are said to support the entire nesting U.S. population of the rare dusky Canada goose.

The risk from the spilled oil was the greater because the sound was enclosed, and therefore the oil slick was continuous. The oil could not break up as it would in the open ocean, so the benzene and other volatile components of the spill, instead of evaporating, soon dissolved in the water to be consumed by zooplankton and other microorganisms at the beginning of the food chain, crucial for the life of the Sound. The enclosure of the oil exposed animals to oil for a longer period of time, and allowed oil to soak deeply into the beaches.

How does oil kill an animal? The diving birds and the sea otters are most at risk. The birds catch their food by plunging into the water, diving to catch their fish, coming to the surface and taking off to fly to their nests. Any weight on their feathers makes that flight impossible. If they get to shore, they try to clean the oil off their feathers with their beaks (preening), thereby ingesting the oil, which is fatal. Since oil makes their insulation (the inner layer of down feathers) matted and useless, most of them freeze to death before they have time to die of poisoning. Of particular concern were two rare birds, yellow-billed loon and merlet, which may be very badly affected (when Exxon applied for permission to dispose of oil-soaked wastes some weeks later, its list of contents to be disposed included twenty *tons* of dead birds). A wildlife photographer counted 650 dead birds on a half-mile of beach on Barren Island. But the observers saw only the birds that managed to struggle to shore. The majority of the dead birds, weighed down by their oil-soaked plumage, sink to the bottom and drown. The sea otters were equally at risk. Unlike the other marine mammals of the area, seals and sea lions, otters have no blubber to keep them warm, but insulate themselves with the air trapped in their thick soft undercoat. Oil destroys their insulation, so only a little on them will quickly freeze them to death in the frigid water. The oil also destroys their eyes, lungs and intestines (when they ingest it from attempting to lick their coats clean).

Alyeska Pipeline Service Company, the consortium of seven oil companies (including Exxon) that actually owns the oil pumped out of the North Slope, had promised, when they were seeking approval for the pipeline, that the operations in Prince William Sound would be the safest in the world. Alyeska's plan, approved by the Alaska DEC, had specified that containment booms and skimming equipment (machinery for mechanically lifting the thick oil off the top of the water into transfer barges, which would take the oil to shore and offload it) would be on the scene of any spill in five hours, with backup equipment (lasers for burning off patches of oil, chemical dispersants) available if the booms and skimmers were inadequate to the job. There was no doubt in anyone's mind that speed was of the essence to contain a big spill. Unfortunately, Alyeska had estimated that a spill of the size of this one could happen only once every 241 years, which made it seem pointless to keep all that equipment around and all those experts on the payroll. In one cost-saving move after another, safety measures were cut back, the contingency plan was trimmed and weakened, specialized crews were disbanded and the only oil-skimming barge (of several promised) was in drydock for repairs at the time of the spill. This failure was particularly bitter for the Alaskans, since the firm and detailed promises of available booms, barges, and crews to handle spills had been what persuaded them to accept the pipeline in the first place. When the weather, which had been balmy for two days after the spill, turned rough, Exxon considered using dispersants (detergents that break up the oil spill) which are made more effective by wave action. Dispersants have problems of their own, however, as we are beginning to discover from their use in the Deepwater Horizon spill in the Gulf of Mexico. They do not really remove the oil from the water. They just sink it a few feet below the surface in an emulsion of oil and detergent (killing all marine life in that area just below the surface); the emulsion ultimately diffuses in the ocean. It kills young fish, and there is no doubt that it would hurt the otters and birds even more (since the detergent would dissolve the natural oils that insulate fur and feathers); the only real good that dispersants do is make the oil spill disappear. In any case, there was nowhere near enough dispersant on hand.

The Exxon Valdez was by no means the only, or the worst, oil-spill incident that has plagued the country. There had been a spill just a few days previously in Valdez harbor. And on March 2, 22 days before the wreck, the Exxon Houston broke loose from her moorings in a storm off Barber's Point, about 15 miles from Waikiki Beach, and went aground on a coral reef about 2000 feet off the west coast of the island of Oahu. Three months later, the nation watched fascinated as three major spills happened in twelve hours: On June 23, 4:40 PM, the World Prodigy struck a reef in Narragansett Bay, RI, apparently because the ship had blundered into unfamiliar waters without waiting for a pilot and repeated the error of the Exxon Valdez, straying to the wrong side of the red channel buoys. 420,000 gallons of Number 2 Fuel Oil flowed into the waters of the bay. At 6:20 PM the same day, an oil barge collided with a tanker in the Houston ship channel, spilling 250,000 gallons of crude oil, about half of which had been cleaned up four days later. At four o'clock the next morning, the Uruguayan tanker *Presidente Rivera* managed to stray from its channel in the Delaware River and hit a rock, spilling 800,000 gallons of Number 6 fuel, rather little of which was recovered. There seems to be something seriously wrong with the entire practice of petroleum transport. In 1988 there had been over 5000 spills involving oil and other toxic substances along our coasts and rivers, which might be considered an "improvement" over the 13,000 a year in the 1970s. Nor did the spills stop. On December 3, 1992, the tanker *Aegean Sea* grounded off the coast of Spain; 23 million gallons of crude oil were lost, most in the fire that followed the explosion, the rest in the sea. In the first week of January, the Liberian-registered tanker *Braer* lost power and went aground, eventually spilling all its 22 million gallons of oil, and severely damaging wildlife–and again, salmon hatcheries. Then the Danish tanker *Maersk Navigator* collided with another tanker on January 21, and its 78 million gallon cargo began to spill into the sea. (Much of that, also, burned off). Then on November 1, 2005, a Hong Kong-flagged cargo ship (*Eider*) ran aground near Antofagasta, spilling heavy bunker fuel and fouling over 7 km of Chilean coastline. Nor do any of these, including the wreck of the Valdez, claim the title of largest: 88 million gallons of crude was lost off Trinidad and Tobago in 1979 when two oil tankers collided; 79 million gallons went into the ocean off Cape Town, South Africa, in 1983.

Tanker and barge accidents account for only two-thirds of the oil that spills accidentally into the waterways each year (an estimated 91 million gallons, for instance, in the period 1980–1986). By far the largest portion of oil that pollutes our waters comes from ordinary runoff, the non-point-source pollution that is so very difficult to stop. The rest comes from pipe ruptures, like the wellhead blowout of the British Petroleum Deepwater Horizon Well in spring 2010 (following upon the 2006 Prudhoe Bay rupture, also owned by BP), or the rupture of an Exxon pipe that filled the Arthur Kill, off Staten Island, with oil in January 1990. Probably the worst spill of all was from the largest act of eco-sabotage known to history: the deliberate destruction of the oil wells of Kuwait by the retreating Iraqi army during the Gulf War in 1991, which released 250 million gallons into the desert and the Persian Gulf. (The BP Deepwater Horizon only released about 170–180 million gallons in all).

Further Reflection: The Reentry of the Ecocentric

Part of the hold that the *Exxon Valdez* has on us is the location of the disaster. Alaska has a place in the American mindset occupied by very few places in the nation as a whole. Alaska, for us, has always been The Wild, the untouched, the final frontier where Nature still rules, untrammeled and free. Some part of us likes that, a lot. With the wreck of the Exxon Valdez, we were suddenly reminded that parts of our universe which we held in high value were under lethal threat from any misstep or complacency in the running of our commercial empires. The Exxon Valdez accident affected us like the photographs of the clubbing of baby seals, or the hunting of wolves by helicopter. Sea otters, even if not "cute" by pet shop standards, are appealing, and their struggles to stay alive touched our hearts. Even if, at some level, we knew such deaths were going on, to see them in Alaska was an especially painful blow. The love of the land, which started off the environmental movement in the U.S., returned as a lead motive for environmental protection.

As that happened, more global environmental threats came to light. For awhile we had worried about the hole in the ozone layer, which raised the possibility of death by unfiltered ultraviolet rays for all living creatures. We got that in hand (we hope) with the Montreal Protocol (1989),[30] which limited the uses of chlorofluorocarbons in refrigerants and consumer products, in an effort to phase them out completely. But now came the threat of global warming, the extensive climate change brought about by the excessive release of "greenhouse gases" (the gases that bring about the "greenhouse effect," warming the earth), especially carbon dioxide from automobiles and factory emissions.[31] After years of denial that such an effect existed, most of the business community recognized the reality of Global Warming, although they were unclear about what to do about it, and how fast that would have to be done.

---

[30] The full Protocol can be found on the United Nations Environmental Programme website: http://www.unep.org/ozone/pdfs/montreal-protocol2000.pdf.

[31] One of the foremost authorities on matters environmental, Gus Speth, then dean of Gifford Pinchot's old school, the Yale School of Forestry—now the Yale School of Forestry and Environmental Studies—brought out a superb book on the subject, *Red Sky at Morning*, Yale, 2004.

Chapter 2
Business Hears the Call

Abstract The positive response of American business to the demands of environmental sustainability is traced, from the earliest cases where technological sophistication made environmentally-friendly products and processes a major goal of manufacturing through to the new emphasis on "sustainable development." New paradigms emerge in the relation of business to the natural environment—biomimicry, the preservation of biodiversity, the orientation to "biophilia," and ultimately William McDonough's replacement of "cradle to grave" with "cradle to cradle."

Keywords Corporations • 3M • Green marketing • Niche marketing • Body works • Ben & Jerry's • Green automobiles • McMansions • Sustainable development • Biomimicry • Biophilia • William McDonough • Cradle to cradle

As the fate of environmental protection had worked out to this point, the major opposition was between corporate business enterprise, with all its claims about providing jobs and stimulating the economy, and the growing environmental movement. The attitude of American business to the natural environment had been until this point one of benevolent exclusion. The corporate leaders loved nature, and encouraged their sons into the Cub Scouts and their wives into the Garden Clubs, but business was business, and the wildflowers don't pay the rent. They had the proof from Adam Smith that the best thing for the greatest number in the long run was to let business pursue profit wherever it could find profit, with no interference from the State. Most of the battles had been won by business, in the past; recently, environmentalists had won more of them and now the future looked dangerous. A new move began circulating through the business system, one that suggested that profits could be made in alliance with the environment. At first, there were only a few pioneers, but slowly a new vision of American business emerged.

Case 11: The Business Case for Environmental Protection: 3M & NYC

Can we make a case that business will profit if it adapts its business practice to protect the environment? The first part of the business case is that environmentally friendly practices limit the impact of crippling hostile regulation, a real danger after Love Canal, Bhopal, and the Exxon Valdez. But the case continues. As early as the end of the 1970s we had begun to collect the iconic "win-win" cases, where measures taken to improve environmental performance within a company turn out to save money. Among the famous tales of cost-lowering environmental measures is 3M's success with its Preventing Pollution Pays (3P) program.[1] Begun in 1975 as an attempt to eliminate sources of pollution, 3P turned out to save money—millions in the first year, nearly a billion by 2005. They did it by restructuring their entire manufacturing process, reformulating products and redesigning the equipment, so that every bit of material that came into the plant was used in the product and not piped out the back door. 3M was really just engaging in good process stewardship, and incidentally proving that environmental progress really does pay.[2] On the public side, New York City, by investing in upstate real estate to surround its water supply, saves itself the $6 billion filtration plant that would be necessary otherwise to ensure the integrity of its water supply. (Or up to $7 billion; all writers use this example, and estimates vary.) With the increase of population in both areas, that beautiful solution may not be feasible much longer; as the value of upstate real estate increases, New York City's payments may not be sufficient. But the principle stands: if we can do something that we have to do cheaper and more efficiently, then the business case is easy to make, with all references to the natural environment only icing on the cake.

Case 12: Green Marketing and the American Consumer

The next part of the case, riding the wave of enthusiasm for all things environmental at the end of the 1970s, came as "green marketing." As R. Edward Freeman points out, the logic of this shade of green is the old-fashioned rule, give the customer what he wants.[3] Consumers turned out to be willing to "buy green," to pay a premium for products that can be seen as beneficial, or not harmful,

---

[1] See John Elkington's account in *Cannibals With Forks: The Triple Bottom Line of 21st Century Business*, pp. 53–54.

[2] A good description of the 3M project can be obtained from Jim Kotsmith, 3P Program Leader, PO Box 33331, 3M Company, St. Paul, MN 55133-3331.

[3] R. Edward Freeman, Jessica Pierce, and Richard H. Dodd, *Environmentalism and the New Logic of Business,* op.cit., p. 14.

to the environment. Corporations who locked into this trend were able to charge premium prices for products whose competitors were spread across the market and competed only in price. Getting out of commodity competition and into a niche market had long been seen as an excellent way of increasing profits; the environment was now a niche. The final stage in this development was the move to organic local food—and the expensive "farmers' markets" that sprouted all over California and New England (and many other places, too). First level examples from this moment would have to include Anita Roddick's Body Works, specializing in "natural" products from the rainforest in her cosmetics and body health compounds, formulated without animal testing. (Cosmetics prior to Roddick had routinely been tested for safety by the "Draize Test," entailing spraying or rubbing the product into the exposed eyes of rabbits, causing pain and sometimes destroying the animal's eyes.) Roddick was stung by criticism that while her products were not developed with animal tests, some of the components she bought were not so innocent. To preserve the brand, she was forced to look *upstream*, to take responsibility for the practices in her supply chain. In this, she and her critics set a precedent that has emerged as controlling at present: each manufacturer shall be responsible not only for his own products, but also for any ethical or environmental problems that attend the origins of his products' components.

Roddick and Body Works were not the only examples of the new trend, nor of the new vigilance upstream. Ben & Jerry's Ice Cream had already fed the supply chain into their marketing: their Rainforest Crunch included Brazil Nuts bought from rainforest cooperatives, their brownie mix ice cream bought brownies from a bakery run by and for the handicapped, and they would only buy cream for their ice cream from cows that had not been treated with rBST—recombinant (artificial) bovine somatotrophin hormones—to increase the milk production.[4] The next move in green marketing had to do with the established industries who suddenly were asked to look *downstream*: you thought you were environmentally friendly, but why are so many of your products in the landfill? Especially with toxic components? Computers were hit hardest. Suddenly, what you do with your part of the operation is not enough: we must look upstream and downstream to evaluate the effects of the enterprise. Not all of this was accepted at once. But current efforts by the computer industry demonstrate that the cradle-to-grave imperative had got some traction.

Green consumption had spread through the economy in the latter part of the twentieth century, and in small ways continues to take its place as an element in the move to sustainability. It is shot through with ironies, most of which are common knowledge. For example, the Organic Food section is the fastest growing corner in the supermarket, even as Wal-Mart becomes the largest retailer

---

[4] Incidentally, the industry is still not happy with the BST fuss; see the Op-Ed piece in The New York Times Friday, June 29, 2007 (Henry I. Miller, Hoover Institution, "Don't Cry Over rBST Milk"), insisting that milk prices would go so high that poor people would no longer be able to buy milk if rBST were taken out of use.

of food; the "fresh and local" movement that crowds the farmers' markets in the affluent suburbs has been completely overrun by the contrary move to the globalized food chain.

Sustainability and the Triple Bottom Line

Through the last quarter of the twentieth century, there was slow and steady progress on the ground toward the business case for environmental protection. First, compliance with regulation, even reluctant, saved money in fines, and could be justified as good business; second, product reformulation and process redesign in some of the more polluting industries could save billions (and incidentally help clean up the environment); third, a public increasingly aware of environmental issues might respond to appeals to buy products made in ways that display sensitivity to environmental issues.

In the 1990s, two major developments transformed our approaches to environmental protection. Until that time, the national (and international) dialogue had focused on regulation for health and safety, with special attention to avoiding pollution of the environment. But the Earth Summit in Rio de Janeiro in 1992 focused the world on the environmental agenda much as Earth Day in 1970 had focused the nation, and gave the world the concept of "sustainable development." The term had first been used in 1980, when the International Union for the Conservation of Nature published its *World Conservation Strategy*. It came into common usage following the 1987 report of the United Nations World Commission on Environment and Development, headed by Gro Harlem Brundtland, Prime Minister of Norway. That Report, entitled *Our Common Future,* defined sustainable development as socio-economic progress that "meets the needs of the present generation without compromising the ability of future generations to meet their own needs." In that form it was picked up and incorporated into Agenda 21 of the Report of the United Nations Conference on Environment and Development (UNCED), a wide-ranging prescription for protecting the environment into the indefinite future.

By the end of the decade, these reports informed the business ethics literature. If the market system is to survive, the corporation must meet "The Triple Bottom Line," a concept advanced by John Elkington in a surprisingly well-distributed book called *Cannibals With Forks: The Triple Bottom Line of 21st Century Business* (1998). Elkington's thesis is that capitalism is not sustainable as capitalism unless it openly adopts environmental goals and strategies, and he provides in that volume a "toolbox" to help businesses get to the point that they need to be.[5] Along the same lines, but more difficult to apply, was *The Natural Step For Business: Wealth, Ecology and the Evolutionary Corporation,* by Brian Nattrass

---

[5] John Elkington, *Cannibals With Forks: The Triple Bottom Line of 21st Century Business*, Gabriola Island: New Society Publishers, 1998.

and Mary Altomare (1999), which also spelled out algorithms for corporations to employ in approaching the desirable condition of environmental sustainability.[6] The point is not just that these were good books with good ideas. The point is that, even while American free enterprise still floated in the backwash of the ideology of deregulation, corporate officers as well as the rest of us, were talking about the *obligation* to meet the standard of environmental sustainability, without intermediate reference to regulation, savings on process or product formulation, or even making a good impression on the environmentalist public. The notion of sustainability was becoming incorporated into the economic psyche of the American citizen, and that is a very good development.

Case 13: The New Industrial Revolution and the Automobile

One of the key works in this development, written by the founders of Rocky Mountain Institute—Paul Hawken, Amory Lovins and Hunter Lovins—was *Natural Capitalism: The New Industrial Revolution,* which also appeared in 1999.[7] It was probably not the first work to point out that there were ways of doing our basic work and living which were a quantum leap beyond current practice in terms of efficiency and environmental friendliness. But by the late 1990s, when it appeared, a new paradigm was needed. Environmentalism seemed to have run out of steam. Industries still lobbied for an end to all regulation, Detroit's automobile makers insisting that the mileage requirements for their automotive fleet could not possibly be increased, or they would surely be bankrupt (and all their workers unemployed), but there was no one to push back. President Ronald Reagan ("Government won't solve your problems; government *is* the problem,") and his administration, over the course of the 1980s, had broken the spirit of regulation. Even with an environmentalist in the executive branch, starting in 1992, very little more was expected from government's angle. But from the point of view of profitable business, quite a bit of progress was possible, and *Natural Capitalism* called on business to take profitable advantage of what was already known. Take the American automobile, for instance, mainstay of the twentieth century economy.

The contemporary automobile, after a century of engineering, is embarrassingly inefficient: Of the energy in the fuel it consumes, at least 80 % is lost, mainly in the engine's heat and exhaust, so that at most only 20 % is actually used to turn the wheels. Of the resulting force, 95 % moves the car, while only 5 % moves the driver, in proportion to their respective weights. Five percent of 20 % is one percent—not a gratifying result from American cars that burn their own weight in gasoline every year.[8]

---

[6] Brian Nattrass and Mary Altomare, *The Natural Step for Business*: *Wealth, Ecology and the Evolutionary Corporation*, Gabriola Island: New Society Publishers, 1999.

[7] *Natural Capitalism: Creating The New Industrial Revolution*, Boston: Little Brown, 1999.

[8] Id. p. 24.

And the situation is getting much worse. As the twenty-first century opened, Americans wanted more cars per person—all newer houses in suburban neighborhoods have three bays in the garage. The cars are getting much larger—the Sport-Utility Vehicles (SUVs), built on the model of trucks—not to mention the Hummers, built on the model of tanks—competed for a growing market in size and weight, consuming much more gasoline than standard cars, projecting an image of intimidation while compiling a frightening record of rollover deaths. This change adds significantly to the greenhouse gases: while the average American car puts an alarmingly high 10,168 pounds of carbon dioxide into the air yearly, the SUV dumps a disgraceful 11,972 pounds into the atmosphere.[9] Meanwhile, American habits of consumption are spreading over the rest of the world, especially China and the developing world, ensuring that any gains in prosperity among those nations will quickly be turned against what remains of the natural environment.

Yet there are powerful indications that the automotive industry can change: that technology is now available to make the American car safer while putting much less of a demand on environmental resources. We might start with the size, weight, and materials of the automobile. Following in the footsteps of Henry Ford, the automakers use steel, which is strong, attractive, and very heavy. A typical six-seat Ford Taurus, for instance, weighs about 3,140 pounds. Using lighter but more expensive materials like aluminum and magnesium, concept cars have been shown that are much lighter. But if the whole metal body is replaced by molded composite materials, embedding carbon, Kevlar, glass, and other fibers in molded plastics, we can get the weight down to about 1,300 pounds. A smaller sedan might weigh as little as 1,000 pounds. The lighter car translates into greater efficiency in several ways: since much less power is needed to accelerate the car, the engine can be smaller and lighter; since much less power is needed to stop the car, the brakes can be lighter; and since there is less weight on the tires, they lose less energy in heat.[10] (As the car becomes lighter, some features, like power steering and power brakes, might become completely unnecessary.) Nor is the light car "less safe" than a heavy one, as anyone who has watched the Indianapolis 500 race will appreciate. The race cars are made of ultralight composites, and despite terrible crashes, the race cars' safety systems usually prevent serious injury to the driver.[11] Anything that can keep a driver safe in a head-on crash at 150 miles per hour should be able to keep driver and passengers safe at road speeds. We know that this is possible: Boeing Corporation has a prototype of a lighter airplane, the 787 (launched on 7/8/07), made largely of composites, that will save massive amounts of fuel. Note that its launch was the most successful in aircraft history, selling hundreds of airplanes on the opening day. Prototypes of extra light "hypercars," that can attain 90 miles per gallon of gasoline, have been with us for decades; why don't we make them?

---

[9] Brian Lavendel, "GreenHouse," *Audubon* March–April 2001, p. 78.

[10] Id. pp. 27–29.

[11] Id. p. 30.

Case 14: The McMansion

The next great unsustainable in the American landscape, target of environmentalist criticism, is generally the American suburban house. The suburbs themselves are targeted for their dependence on the car; see above. But the freestanding house on the large lot in the suburbs is an independent assault on the environment. On the average, the American single-family home emits 16,522.3 pounds of carbon dioxide from its use of electricity generated in plants that burn fossil fuels. If it burns oil in the furnace, it costs 17,622.0 pounds of carbon dioxide to heat the house (natural gas lowers the amount to 11,094.3 pounds).[12] That's about 30,000 lbs, or 15 short tons, per family. And it's getting worse. The house that now sells best on the American market is the "starter castle," or "McMansion," the outsized house with a ballroom-sized two-storey front hall, an enormous eat-in kitchen plus breakfast room, formal dining room, family room, perfect for entertaining, with room for the two children, the corgis, and the *au pair*. All this room will require more fuel to light, to heat, and to supply electricity for the entertainment centers, computers, fax machines and kitchen appliances. Nor does this total include the amount of gasoline burned in mowing the lawn with power mowers, or blowing leaves into a pile with power-driven leaf blowers so that they can be vacuumed into a truck and driven to the dump.

Our houses, like our cars, like our diets, seem to be controlled by the need to "supersize." While we supersized cheeseburgers and soft drinks, we were supersizing our houses, and going way into mortgage debt in order to do it; there is a strong link among the bubbles—financial, real estate, all the material dreams that popped in the collapse of 2008–2009. Yet houses at least can be sustainable; buildings have been designed (and in some places built) that use much less energy (through superinsulation, efficient heating plants, and double windows). Where photovoltaic panels already available are attached, they can, like a tree, create more energy than they use.[13] Much more can be realized from the careful design of industrial and office buildings. Universities might very well be the leaders in this development—we make very public buildings, with a lot of freedom to design what we want. There are now new institutions to guide and institutionalize environmental design: Leeds certification, for instance, providing a target and verification for designing and building green buildings. From just the progress reported in 1999, it seems that we could take the entire industrial system of the developed world, wring out the waste, redesign the basics, and make a profit in the process. The technology is there; it is the mind of the technologist that needs changing.

---

[12] Lavendel, op.cit., p. 78.

[13] See Tom Zeller, Jr., "Can We Build in a Brighter Shade of Green?" *The New York Times*, Sunday, September 26, 2010, first Business Page.

Case 15: Biomimicry and Beyond

That mind, in the recent past, even before the supersize dysfunction, has been slow to recognize the extraordinary technological sophistication of the natural world. For instance, all of our clean water (not just New York City's) is ultimately the product of a water filtration governed entirely by natural processes; our medicines are derived, often in one step, from the incredible variety of plants in the natural biodiversity of the threatened rainforests.[14] Nature's riches and nature's services do not often make the news until we manage to cripple them somehow; in recent years, we have heard repeated alarms over "colony collapse disorder," destroying our honeybees, the pollinators on which many of our fruit, seed and nut crops depend.[15] If we lost the bees, we have no idea how we might replace them. Could we take Patagonia (an outdoor clothing company that models all its products on nature) rather than the chemicals company 3M as the model, and redesign our activities to conform with nature? Should we adopt the philosophical orientation of belonging to, and working with, the natural world, instead of imposing our designs upon it, we might discover wholly new ways of doing what we wanted to do and doing it profitably. Could Nature teach us how to do things better? We make Kevlar, one of the strongest fabrics we know (used for sails in competitive yacht racing, body armor for the military), in boiling sulfuric acid. Spider silk, which is stronger, is made at room temperature with only natural materials. Can we learn to do that? A new field, Biomimicry, grew up in efforts to find more efficient ways to make and do what we want.[16] To recap our first question, why do we want to preserve the environment? *Because Nature contains resources, and models processes, that we need to use and to know in order to carry on our economic activities.*

The revolutionary approach to industry was not new. As early as 1970, Barry Commoner had pointed out that there was no such place as "away," and that therefore we really didn't throw anything there; what we thought we threw there actually stayed with us, rotting and polluting.[17] What can we do with waste, waste that

---

[14] For a start at accounting these services, see Gretchen Daily, ed. *Nature's Services: Societal Dependence on Natural Ecosystems*, Washington, DC: Island Press, 1997; also Robert Costanza et al., "The Value of the World's Ecosystems and Natural Capital," *Nature* 387:253-260, 1997.

[15] To quote the entomologists of Cornell University Agricultural School, "Recent reports in the news have highlighted a dramatic loss of honey bee colonies in as many as 24 states, and the number is growing. Honey bees are a critical player in the production of many fruit, vegetable and seed crops grown throughout the country, adding between $8 and $12 billion worth of value to US agriculture each year. Substantial losses, such as are currently being experienced, pose a serious threat to crops that rely on bees for pollination and portend diminished profits for growers and higher prices for consumers at the supermarket." http://www.entomology.cornell.edu/ithacacampus/articles/beecolonycollapse.html.

[16] Janine M. Benyus, *Biomimicry: Innovation Inspired by Nature*, New York: HarperCollins 2002 (William Morrow 1997).

[17] Barry Commoner, *The Closing Circle: Nature, Man and Technology*, New York, Knopf, 1971.

cannot be thrown "away"? William McDonough, the revolutionary green architect, proclaimed a new paradigm for industry, "waste is food": The standard model of industrial "throughput," resources entering at one end of a factory, product exiting at the other end, waste departing to someplace "away" through a pipe or in a truck, had to be abandoned. Adopting the perspective of biomimicry, he pointed out that in the natural world, nothing is lost; every bit of waste from one natural process becomes the raw material for the next life. Our 1970s intolerance for wastes that persist in our environment and eventually endanger our health has led to the requirement that we shepherd our products to some safe end, "cradle to grave." McDonough's new prescription insisted that we follow them "cradle to cradle"—from the creation of one product to the resources necessary for the next.[18]

Reflection: The Philosophers

What contribution to the ethics of the environment came from the philosophical community? "The owl of Minerva flies only at dusk," we philosophers say of our field, or in other words, philosophers as usual paid no attention to what was going on in the world while it was going on, and had to play a catch-up ball of Analysis and Reflection at the end. The battle is nearly over before the philosophers show up, and they are not the ones that lead the next phase.

They started from the basics, adopting the orientation of Jeremy Bentham. Consider, they suggested, the wild animals, living before and away from humans. They too have values, along the lines of all humans. They seek the necessaries of life, food, warmth, and the opportunity to reproduce, and for them life is good when that necessity is satisfied; they seek pleasure and avoid pain; they have lives that can go better or worse for them. As Holmes Rolston III put it,

> There is no better evidence of non-human values and valuers than spontaneous wild life, born free and on its own. Animals hunt and howl, find shelter, seek out their habitats and mates, care for their young, and flee from threats. They suffer injury and lick their wounds. Animals maintain a valued self-identity as they cope through the world. They defend their own lives because they have a good of their own. There is somebody there behind the fur or feathers.[19]

Animals value their own lives, and things in their world, their own lives intrinsically and their resources instrumentally. If those facts are enough to recommend

---

[18] William McDonough and Michael Braungart, *Cradle to Cradle: Remaking the Way We Make Things*, Farrar Straus Giroux 2005. One of the more prominent adherents of this new wave is Ray Anderson, founder and CEO of Interface, Inc., which has developed furnishings that are completely sustainable—using only dyes that are biodegradable and non-toxic, materials that can be recycled, and minimizing the use of fossil fuels. See his *Mid-Course Correction*, Distributed by Chelsea Green in White River Junction, Vermont.

[19] Holmes Rolston III. "Environmental Ethics", in *The Blackwell Companion to Philosophy*, 2nd ed., ed. Nicholas Bunnin and E.P. Tsui-James, Oxford: Blackwell Publishing, 2003, pp. 517-530.

a valuing of human life and flourishing in our world, it should be enough for the animals. So there can and ought to be an animal welfare ethic; or, some prefer to say, an animal rights ethic.

How can we deny that the same moral imperatives apply to them as to us? Consider Jeremy Bentham's conclusions:

> The day may come when the rest of the animal creation may acquire those rights which never could have been witholden from them but by the hand of tyranny... What... is it that should trace the insuperable line [between human and brute]? Is it the faculty of reason, or perhaps the faculty of discourse? But a full-grown horse or dog is beyond comparison a more rational, as well as a more conversable animal, than in infant of a day, or a week, or even a month, old. But suppose they were otherwise, what would it avail? The question is not, Can they reason? Nor Can they *talk?* But, *Can they suffer?*[20]

Bentham was not taken seriously, then or now, because of our pervasive certainty that no matter how logical his argument, it was perfectly clear that humans were above nature, and possessed a value all their own. Yet the same kind of reasoning that made "beneficence" a moral imperative leads to the conclusions that over this range at least, the higher animals at least are very like us, and if we have in some sense a "right" to life and to the means to preserve it, then so do they. (Those who argue for some "special creation" for humans should remember that chimpanzees were created, by the same God, with 98.5 % of their DNA identical with ours. And don't get too excited about that 1.5 %; it has to do with our ability to lock our knees.)

Humans are in radical disagreement on the status of even the highest of the animals. We may find it easy to identify with animals like ourselves (the great apes, for instance), but others of our species treat them as "just animals" and cheerfully shoot them for the cooking pot. As far as that goes, humans have practiced cannibalism into modern times, simply asserting that people of other tribes were the equivalent of "just animals." The numerous "nature" programs on television, portraying animals sympathetically, testify at several levels to a human propensity to identify with animals: that TV producers with a finger on the public pulse think such programs will sell, that naturalists, photographers and narrators dedicate their lives to studying and understanding and representing the beasts they portray, and that large numbers of viewers find such portrayals attractive.

Shall we grant "rights" to the animals? There is a powerful movement within philosophy to accord basic rights to at least the higher animals. What rights would be appropriate to accord? Most obviously, the right to life—at least, the right not to be hunted or slaughtered for food, and there goes the larger part of the world's diet, and a huge profitable industry in the U.S. It's the kind of conclusion that environmentalists love, for whether or not you are a vegetarian, you'll probably recognize that eating closer to the land will save the environment.

Then shall we grant rights to the plants? Ecocentric ethics requires respect toward all living things, not only the chimpanzees and horses and dogs of our

---

[20] *Introduction to the Principles of Morals and Legislation*, Chap. XVII, note 1. Hafner Library of Classics, #6, Hafner Pub. Co., New York, 1948, p. 311.

experience, but Monarch butterflies and the California redwoods. The vast majority of biological organisms are left out of the "animal rights" agenda—the snails, insects, microbes, and plants, for instance. As Holmes Rolston reminds us, "Over 96 per cent of species are invertebrates or plants; only a tiny fraction of individual organisms are sentient animals."[21] A plant is certainly a life-form, arising spontaneously in nature, responding to stimuli (turning to face the sun, for instance), able to sense certain events (like the attack of a predator) and within limits respond—often by producing a poison to get rid of it. It eats, digests, excretes waste, grows, repairs wounds, maintains its identity and reproduces its kind. Clearly its state of existence, its life, is of value to it, however inarticulate it may be in expressing that value.

An ethical problem haunts the extension of rights to non-human species: it is irresistible to the philosophers, to lump this move with the civil rights movement generally. First only white males were accorded rights; then nonwhite males; then females; then chimpanzees; then horses; then dogwoods; etc. We steadily expand the bounds of the rights holders, becoming more inclusive with each move. All of this is good. But there has to be a difference between including human beings, formerly enslaved, in the privileged group, and including even the higher animals. Humans should never have been enslaved, should never have held a status less than full humanity; their inclusion was dictated by justice. There are excellent reasons to exclude animals of all kinds from full rights holder status—just no reason to inflict pain on them. That's a difference for which we have yet to account.

Value does not lie only in the satisfaction of human preferences or desires. Value lies in life, in the whole interrelated world, in the entire realm of what we call Nature. We can divide it as we like, as its protection seems most convenient. Saving the individual organism is almost never first priority (exceptions for the last remaining tortoise of its species). The smallest unit worthy of protection is probably the ecosystem, a local community of rocks, streams, plants, and animals, living in interrelationship, interdependent, destruction of any part of which will severely compromise the rest. The ecosystem is an organism in itself, like our bodies, and any insult received by one sector will be felt throughout. It has been suggested that the Endangered Species Act should be supplemented by the Endangered Habitat Act, since it makes little sense to go to desperate measures to protect the last members of a species, if we continue to destroy the habitat that the species needs. Ultimately, the unit of protection is the biosphere itself, the entire realm of living beings.

Biodiversity

But must we pay attention to how many different kinds of organisms we organize to protect? There is no doubt that "biodiversity" is a value in environmental philosophy. We referred to it briefly among the critiques of Norman

---

[21] Rolston op.cit. p. 521.

Borlaug—protection of the variety of species is one of the goals of environmental protection. But why? Edward O. Wilson, the Harvard biologist who pioneered this subject, suggests the following:

> Biological diversity—"biodiversity" in the new parlance—is the key to the maintenance of the world as we know it. Life in a local site struck down by a passing storm springs back quickly because enough diversity still exists.
>
> Opportunistic species evolved for just such an occasion rush into fill the spaces. They entrain the succession that circles back to something resembling the original state of the environment. This is the assembly of life that took a billion years to evolve. It has eaten the storms—folded them into its genes—and created the world that created us. It holds the world steady....[22]

The complexity of the system, that gives it its resilience and versatility, can be found in the smallest and most basic samples of biological life, in a handful of soil scooped up from the woods.

> The black earth is alive with a riot of algae, fungi, nematodes, mites, springtails, enchytraeid worms, thousands of species of bacteria. The handful may be only a tiny fragment if one ecosystem, but because of the genetic codes of its residents it holds more order than can be found on the surfaces of all the planets combined. It is a sample of the living force that runs the earth—and will continue to do so with or without us.[23]

He goes on to cite the rate of extinction of species occurring right now: hundreds, thousands of times higher than before humans arrived on earth. "They cannot be balanced by new evolution in any period of time that has meaning for the human race." But does it matter? Wilson restates the classic arguments for attention to biodiversity:

> [If we allow many of the species of the earth to disappear,] New sources of scientific information will be lost. Vast potential biological wealth will be destroyed. Still undeveloped medicines, crops, pharmaceuticals, timber, fibers, pulp, soil-restoring vegetation, petroleum substitutes, and other products and amenities will never come to light. It is fashionable in some quarters to wave aside the small and obscure, the bugs and weeds, forgetting that an obscure moth from Latin America saved Australia's pastureland from overgrowth by cactus, that the rosy periwinkle provided the cure for Hodgkin's disease and childhood lymphocytic leukemia, that the bark of the Pacific Yew offers hope for victims of ovarian and breast cancer, that a chemical from the saliva of leeches dissolves blood clots during surgery, and so on down a roster already grown long and illustrious despite the limited research addressed to it.[24]

Most of Wilson's argument can be summed up as a particularly profound prudence: over a large range, we don't know what species really do in the world, either for us directly, or in their role in their own ecosystems. We dare not fool with systems we do not understand. "We should judge every scrap of biodiversity while we learn to use it and come to understand what it means to humanity. We should not knowingly allow any species or race to go extinct. And let us go

---

[22] E. O. Wilson, *The Diversity of Life*, Cambridge, MA: Harvard University Press, 1992, p. 15.

[23] *Ibid.* p. 24.

[24] Loc.cit.

beyond mere salvage to begin the restoration of natural environments, in order to enlarge wild populations and stanch the hemorrhaging of biological wealth."[25]

Case 16: Biophilia? Circling Toward Home

Many of the most interesting new imperatives for American society were put there by you and me. The "green marketing" literature is all about our taste in consumption, possibly a passing fad. But that literature has been supplemented by some interesting studies that suggested that the link between green life, plants and woodlands, might be considerably stronger than we had thought. The phenomenon known as "biophilia," a natural bond with all living things that guides our choices in life whether or not we are aware of it, may not be scientifically demonstrable, but it seems to account for the fact that we heal from illness better if we can see out of doors, and that we thrive when we are surrounded by living things—green ones and furry ones. Biophilia may very well underlie the appeal of green marketing, and the new awareness of the consumer.

More importantly, it suggests that at a near distance, humans recognize their kinship with nature, and are entirely capable of bonding with, loving, parts of the natural world beyond the human—and beyond the personal back yard, the local park and the community garden, which are already infused with human meanings and symbolic importance.

Reflection: Deep Ecology and the Justification of Monkey Wrenching

In 1973 a new initiative arose in environmental ethics, from a paper by Arne Naess—"The Shallow and the Deep, Long-Range Ecology Movement."[26] The essential distinction referenced in the title of the paper is between anthropocentric environmentalism, preserving the environment for the sake of human safety, welfare or pleasure, and ecocentric environmentalism, preserving the environment for its own sake. In subsequent discussions, it has often been difficult to pin down just what Deep Ecology's core beliefs are, but that initial distinction remains: Nature exists for its own sake, and humans are, and must be, no more than part of Nature. Where human interests oppose the interests of the ecosystem or the larger ecosphere, our tendency is to let the human interests trump the environmental interests; in deep ecology, the resolution is always in favor of nature. Humans must

---

[25] Loc.cit.

[26] Arne Naess, "The Shallow and the Deep, Long-Range, Ecology Movement: A Summary," *Inquiry* 16, p. 95.

regard themselves as organs within the larger body of the earth (or of Gaia, the entire ecosphere of the earth, named for the goddess of the earth, in some formulations). The turbulent history of the earth has resulted in significant die-offs of many of its species, without completely losing the species (usually), and without losing the resilience of the ecosystem, which can continue to nurture other species in their turn. We apparently have found these die-offs acceptable, even when we ourselves have caused them. The deep ecologist will simply point out that what would do the earth the most good right now would be a significant die-off of human beings. (There may have been some in the past, accounting for the fact that humans lived in harmony with the earth's limitations for several million years at least.) One reason that "deep ecology" has been difficult to pin down is that it is disorienting to recommend that 90 % of humans are surplus, and damaging to the welfare of the whole, without going on to specify what kind of "Second Amendment Remedy," as we now call such proposals, it would be reasonable to adopt to address this disproportion. The similarity of humans to a cancer in the earth has been acknowledged, see below; the remedy for cancer is to destroy it, or at least enough of it so it does not threaten its host. The discussion of such remedies, where human beings are concerned, is not tolerated in civilized society, and the deep ecologists are civilized. But such thinking did lead to a new environmental tactic in the woods of the Pacific Northwest that draws on the fundamental convictions of Deep Ecology. Seeing the powerful machinery of the logging companies chewing up the redwood forests, even when laws and court decisions forbade their activities, some of the environmentalists took up "monkey-wrenching," disabling the machinery (without hurting the loggers), throwing a monkey wrench into the next day's work. It was against the law, of course, to damage property; but often the loggers were working in defiance of the law or of court decisions, so the young vandals did not feel as guilty as they might have otherwise. The lumber companies hired better security guards, and the practice faded away (the logging did not, and continues to this day). But a profound ethical question remains: in a nation ruled by law, what law takes precedence in such cases? We may agree that to save human life, it is permissible to damage property. But to save the trees?

Chapter 3
Coming to Value Nature

Abstract This chapter steps back to consider the many orientations to the natural environment, aiming not at praising one or the other, but at permitting an analytic understanding of the ethical differences among them. Some directions that corporations and citizens might take to continue the progress noted in the last chapter are suggested.

Keywords Climate change • Overpopulation • Depletion of resources • Pollution • Loss of species • Energy problems • Seventh generation • Attitudes toward the natural world

Man as a Plain Citizen of the World: Return to Aldo Leopold

Long ago, Humans lived in, on, and dependent upon, the natural world, and did reasonably well. Most notable about that period was not primarily its difficulty, but its stability—for thousands, yea millions of years, humans flourished without changing the environment in any permanent way. That is the ideal to which we aspire to return—not living in caves, but living in a way that allows us to use our fair share of the earth's resources without disadvantaging the other species.

How is this so, that human activities are impacting the ecosystems of the earth, in such a way as to cause alarm and call us to action to bring a halt to it? In one memorable description, humans themselves in all their activities have been likened to **cancer—humanity, the cancer of the biosphere**. Let's see how that might work.

Sometimes cancer simply changes the metabolism of the whole body. Similarly, the first and most far-reaching environmental offense, and the one most in the headlines, is **Climate Change**, the gradual warming of the world through the effects of human activity. We know how it takes place. Some of the heat received from the sun is not absorbed in the earth but is reflected back into space. Our canopy of atmosphere, especially the gas CO_2, Carbon Dioxide, product of the respiration of all living beings, catches some of that reflection and sends it back down to earth as warmth. That is why life is possible; on planets without atmosphere, the sun may shine down brilliantly, but none of that heat will be retained. Because of what has been called the "problem" of the greenhouse effect, we and all life are alive. The problem arises, of course, when we get too much of a good thing. When carbon dioxide builds up in the atmosphere, due primarily to the burning of fossil fuel, the whole earth warms up. In fact atmospheric concentrations of carbon dioxide have increased and are still increasing; people released 6.44 billion tons of carbon into the atmosphere in 2002, a 1 % increase over the year before, bringing atmospheric carbon dioxide to 372.9 parts per million by volume in 2002. (It was about 320 parts per million in 1960.) And global temperatures, famously, have increased: the average global temperature was about 13.8 °C in 1880, when they started keeping records, but the average temperature for the twentieth century was 15.6 °C (60.1 °F), and August 2012 was 0.62 °C (1.12 °F) above that.[1] The polar icecaps are shrinking, losing half a million square miles between 2005 and 2007[2]; after 2000 years of little change, the sea level is rising precipitously (about 6 inches between 1870 and 1990, but another 2 inches between 1990 and 2008, threatening many small island nations with termination),[3] and we seem to be on our way to the earth of two and a half million years ago, when the oceans nearly swamped the earth. Who should take responsibility for the increase in global warming?

The second offense, and the way we recognize most cancers, is **Overpopulation,** the simple crime of proliferation, overgrowth, a mass of a certain kind of cells, more than there should be for the sake of the body. If it does nothing else, this mass of one kind of cells displaces other organs, blood vessels, or in the case of the environment, habitats of other species. There's too much of it for the health of the whole.

The population figures are well known. It took until the modern period for the human race to achieve its first billion on earth; we reached 2 billion in the 1930s, 3 billion in 1960, 4 billion in 1976, 5 billion in 1989, 6 billion in 1999 and 7 billion in 2011. That is a very rapid increase, achieved primarily by lowering the human death rate (for we aren't having that many more children). At this rate, we expect the eighth billion by 2025. But the rate has moderated, due to a further lowering of the fertility rate, and the United Nations now estimates that there will be only 8.9 billion people on earth in 2050, not 9.3 as previously thought. That's a very large "only"! Nor are they very happy people: according to World Bank Development Indicators (2008), about 80 % of the world's population (outside the developed countries) live on less than $2/day.[4]

---

[1] From the National Climate Data Center (NCDC), part of the National Oceanic and Atmospheric Administration (NOAA), website accessed November 13, 2012.

[2] *Ibid.*

[3] *Ibid.*

[4] World Bank Development Indicators 2008, from Global Issues website, accessed November 13, 2012.

Actually, the population figure has to be modified to take into account that some of those masses of cells are a lot more dangerous (malignant?) than others. For the impact a human makes on the earth is not a simple function, one human one impact: the affluent humans, who consume many more units of a given resource than the poor ones, have a much larger impact. Further, where technology aids the consumption, that technology adds its own environmental degradation and pollution per unit of resource consumed. (Example: An American uses three times the amount of wood used by a typical Indian, simply in consumption of chairs, tables, houses, and paper pulp. But even per unit of wood used, the American consumes more than the Indian, for cutting down a tree with a simple axe produces no pollution or destruction of the soil, while felling the tree with a chainsaw causes several types of pollution, and dragging it out of the forest with a bulldozer does enormous amounts of damage to the soil.) So we can draw up a simple equation (following Ehrlich) to summarize the environmental impact of population:

"P (population) × A (affluence, consumption) × T (technology) = I (impact)."

The implications of the equation are disquieting: even though the major increases in population are happening in the less developed world, it is the developed world, especially the United States, that is causing the greatest environmental impact.

If overpopulation is the second offense, the third is **Resource Depletion**, stemming from our monopolization of the resources of the body or the earth. The reason we lose weight when we have cancer is that the tumor commands the body's resources for itself, even spurring the growth of new arteries to direct blood to itself at the expense of the other organs. Similarly humans, instead of adjusting their choices out of consideration of the needs of the other life on earth, crowd into whatever ecosystem seems appropriate to their needs, and proliferate without regard to allocation of scarce resources. (To be more precise, the developed world and its market-based economic system appropriate the resources of the world; the human inhabitants of the developing world fare no better than the rest of Nature.) One way of expressing the human impact on the biosphere is by calculating the fraction of the net primary production of the earth that humans appropriate. Net primary production is the energy left in the biosphere after subtracting the respiration of primary producers (photosynthesizing plants) from the total amount of energy (mostly solar energy) fixed biologically. Net primary production is, in effect, the total food resource on earth. Humans, only one among the millions of animal species on the earth, are already using 40 % of it—more, on some calculations.[5] That's a lot. Since 1950, the use of lumber has tripled, the use of paper has increased sixfold, the fish catch has increased nearly fivefold, grain consumption has tripled, and the burning of fossil fuel, a non-renewable resource laid down as

---

[5] Peter Vitousek, Paul R. Ehrlich, Anne H. Ehrlich and Pamela Matson, "Human Appropriation of the Products of Photosynthesis," *BioScience* 36 (1986).

part of earth's capital millennia ago, has quadrupled.[6] We know that this consumption is impossible to sustain, even if only humans needed the product; but we are taking from every other life form as well.

The next major category of problems is **Pollution**: of air, water, and the earth itself. When a tumor grows, it gives off waste products, toxins that may destroy the ability of the liver to function. When people expand, they give off more waste products too. Air pollution includes the "greenhouse effect," that causes global warming, above. Another cross-border complaint comes from the impact, in Canada, from our Midwest power plants. You know the problem: the tall stacks of the power plants route the smoke up above its surroundings, and successfully keep the pollutants from the homes and businesses of the citizens of the local towns who control its permissions to operate (and pollute). The smoke contains nitrogen oxides (NO_2, NO_3, generally summarized as "NO_X") and sulfuric oxides (SO_X), that are picked up by the rain and transformed into liquid acid (nitric acid, HNO_3, and sulfuric acid, H_2SO_4) that change the chemical composition of the waters wherever the rain falls. Acid rain, or acid deposition, is charged in the United States with killing fish (and everything else) in the lakes of the Adirondack Mountains in New York, and killing the trees of the highest mountains in the Eastern mountain ranges—the Green Mountains of Vermont and the Smoky Mountains in North Carolina, in particular. Acid rain had first been seen as a problem in Europe, where the famous Black Forest showed signs of failure in the late twentieth century. It is becoming a huge problem in China (where the rain occasionally approaches the acidity of vinegar), South America and parts of Africa. Air pollution also comes in the form of little specks, particulate matter, that invade the lungs. The burning of diesel fuel, especially in badly tuned engines, creates a constant stream of such particles. We do not know what other harm it does in its incomplete burning, but we know that it is a major trigger for asthma, especially in children. At least the emissions from smokestacks and fixed engines is locatable, point source emissions that we can find, should we want to control them; more troublesome are the non-point source emissions that come from moving targets like cars and trucks, rain runoff from the lands around the watercourses, and the oil spread through the waters from motorized recreational watercraft.

Another problem caused by pollution of the air came to light in the 1980s. Suddenly the stratospheric ozone layer, the layer of O_3 that shields us from the worst of the sun's rays, was developing holes, especially over Antarctica. How had this happened? For the first time (after the usual bouts of denial!), we were able to identify a single man-made mechanism of environmental deterioration: the chlorofluorocarbons that we had been using as refrigerants and propellants in our spray cans had been attacking the ozone layer, gumming up the continual flow of changes from ionic to molecular oxygen and back again, that was protecting us from ultraviolet radiation. If the deterioration continued, that radiation could scrub the earth clean of life. Once the mechanisms had been identified, the

---

[6] Miller, op.cit. p. 8.

major manufacturing nations got together in Montreal and agreed to eliminate chlorofluorocarbons from their industrial processes, substituting other compounds. (The agreement was signed in 1987 and went into effect in 1989.) The Montreal Protocol is one of the few international environmental agreements that seems to be effective, for several reasons: only developed nations were manufacturing the offending chlorofluorocarbons, the companies engaged in the manufacture had backup products that they could put on the market, and a model compliance method was developed. In short, the ozone depletion problem, compared to the climate change problem, was easy.

The atmosphere is not the only recipient of the waste products of our enterprises. The waters—creek, rivers, oceans—have been our industrial sinks and sewers for about two centuries. Among the worst of the water polluters are pathogens, germs, microbial life that causes diseases in humans if humans drink the water. These are often introduced in human and animal manure. Strictly speaking, the problem is only one of public health, if the rest of the ecosystem is not harmed by the microbes, but it is a serious problem: some 4 million children a year die of diarrhea caused by pathogens in the water supply, according to UNICEF, and those who recover are seriously weakened by malnutrition, vulnerable to many other diseases.[7] Pathogens at least die and disappear in the ordinary course of events. Persistent chemicals like polychlorinated biphenyls (PCBs) do not die, at least not in the real time of human use. These chemicals were discovered and manufactured all through the mid-twentieth century, valued as insulators and lubricators for machinery precisely because they reacted with nothing, never changed, and never broke down—once they were anyplace, they were there, for all practical purposes, to stay. Predictably, tons of them ended up in the waterways, most notoriously in the Hudson River in New York, near the old and leaky General Electric plant that made them. PCBs exhibit the problem peculiar to many persistent organic pollutants (POPs): they are not soluble in water, but are soluble in fat, so when fish ingest them with food, they accumulate in the fish's body fat. When the birds eat the fish, they accumulate in the bird, more PCBs with every fish meal, and there they stay. If we eat the fish, the same thing happens to us; PCBs gather in the fat and in all body fluids that contain fat—for instance, the breast milk of nursing mothers. They have been linked to cancer; they eventually wreak havoc on the immune system, the hormonal system and the reproductive system; and we don't know how to get rid of them.[8]

Even good nourishing fertilizer pollutes the water. When our bodies of water, fresh (lakes and rivers) or salt, accumulate too many nitrogen and phosphorus compounds, they undergo "eutrophication" (literally, "change for the good"), becoming too rich with nutrients for their ecosystems. The new level of nutrition

---

[7] UNICEF website, accessed November 13, 2012.

[8] US Environmental Protection Agency (EPA) website on "Health Effects of Polychlorinated Biphenyls" accessed November 13, 2012.

encourages large blooms of algae, which are decomposed when they die by aerobic bacteria—bacteria that use oxygen. The oxygen supply in the water plummets, and the living things, fish and shellfish, that depend on the oxygen, must leave or die.

The land, too, accumulates pollution. Most visible on the horizon are the "landfills," the enormous dumps in which we put our solid waste. Solid waste is simply any unwanted or discarded material that is not a liquid or a gas. The United States generates about 10 billion metric tons (11 billion of our tons) every single year— about 44 tons per person. With only 4.6 % of the world's population, we manage to produce a third of its solid waste. Before consumers start feeling guilty about that, it should be pointed out that almost 99 % of that waste comes from mining, oil and natural gas production, agriculture and other industrial activities. Mining waste is the worst offender.[9] Attempts to reduce the reducible portion of this enormous pile of waste (and that part includes none of the mining waste) center on alternatives like compacting it and burning it as fuel for the waste treatment process. Recycling might be more effective. Possibly more serious than just the sheer volume of waste is the toxicity of some of it. Toxic, or "hazardous" waste is defined as any waste that is corrosive (tends to eat the tanks you keep it in), reactive, explosive, easily inflammable, or contains one or more of 39 defined toxins— carcinogens, mutagens, and the like. The list contains a good many pesticides, solvents and paint strippers, but does not extend to the exotic wastes generated by special industries—the radioactive wastes from nuclear power plants, for instance, and the slag from mining operations. These, also, have to be dealt with by systems that may not be fully prepared for them.

After the poisoning of air, water and earth, the next most serious problem is **Species Extinction**, the rapid loss of the biodiversity of the world. As cancer quietly displaces and destroys parts of the body that it consumes, so the human race seems to be displacing many too many animals and plants that have been on the earth for more millennia than we have been. Most species loss is attributed to habitat destruction. About 50 % of all species on earth, for instance, are endemic to the tropical rainforests (that is, they are found nowhere else.) According to the World Resources Institute, 7–8 % of tropical forest species will become extinct each decade at the present rate of forest loss and disruption. That's about 100 species per day. Since 1950, almost a third of all tropical forests existing then have been cut down, the land changed to other uses; we have very little idea of how many species we have already lost in that wastage.

The sixth and last complaint in this sample of the environmental abuse is **Energy Waste**, the systematic abuse of energy worldwide. Outside of the small contributions of geothermal and nuclear energy, we have only one source of energy, the sun. The sun not only gives us warmth and light, it is virtually the only source we have for the energy to run the world. For most of the developing

---

[9] Miller, op.cit., p. 579.

countries of the world, all cooking and heating is done with wood—the energy of the sun transformed by photosynthesis and other living processes into long organic molecules and stored in trees and bushes. Wood is renewable where population pressure is not great; unfortunately, the poorest areas of the world have serious problems with population. At the other end of the economic spectrum, the United States, with only 4.6 % of the world's population, uses 24 % of the world's commercial energy, primarily from fossil fuels—the energy of the sun transformed by photosynthesis into organic molecules (in "primary production," described above), then compressed within the changing earth into pure carbon (coal) or condensed hydrocarbons (oil and gas), burned for heat, light, electricity and transportation. India, with 17 % of the world's population, uses only about 3 % of the world's commercial energy. 84 % of all commercial (fossil fuel and nuclear) energy is wasted. 41 % of the energy is necessarily lost; by the second law of thermodynamics, no transformation of energy from one form to another can be completely efficient. But 43 % is wasted unnecessarily, "mostly by using fuel-wasting motor vehicles, furnaces and other devices, and by living and working in leaky, poorly insulated, poorly designed buildings."[10] According to energy expert Amory Lovins, energy waste in this country alone comes to over $300 billion per year—an average of $570,000 per minute. But the wastage is not confined to the developed world: the open fires on which the poorer residents of the developing world do their cooking lose much more heat than they retain. Any worthwhile environmental policy aimed at reducing this waste will have to address every facet of the human use of energy, and the problem promises to be monstrously complex.

The list above only scratches the surface of environmental problems that have to be addressed. It's a short list, and easy to remember: just think cancer, and **CORPSE,** for the problems identified.

Climate Change,
Overpopulation,
Resource Depletion,
Pollution,
Species Extinction, and
Energy Waste

The metaphors are compelling, but a metaphor is not an argument. In this chapter we will develop the orientations available to our culture toward the natural environment, and, very briefly, the Environmentalist point of view, the orientation of the one who will argue, vote, support organizations, March, maybe even lie down in front of a bulldozer, to defend the natural environment against activities of government or business.

---

[10] Miller, op.cit. p. 397.

Ethics Focused on the Natural Environment

The Iroquois Confederation, in convention, adopted the following principle as commanded by their Great Law:

> In all of your deliberations in the Confederate Council, in your efforts at law making, in all your official acts, self interest shall be cast into oblivion. Cast not over your shoulder behind you the warnings of the nephews and nieces should they chide you for any error or wrong you may do, but return to the way of the Great Law which is just and right. Look and listen for the welfare of the whole people and have always in view not only the present but also the coming generations, even those whose faces are yet beneath the surface of the ground—the unborn of the future Nation.

As Oren Lyons, Chief of the Onondaga Nation, wrote: "We are looking ahead, as is one of the first mandates given us as chiefs, to make sure and to make every decision that we make relate to the welfare and well-being of the seventh generation to come… What about the seventh generation? Where are you taking them? What will they have?"[11]

From this wisdom—the wisdom of those displaced and very nearly exterminated by the coming of our forebears to the Western Hemisphere—environmentalists have adopted a simple test for the permissibility of any new technology, change in customs or in hunting patterns: will the choice we make today seem good from the perspective of our descendants seven generations from now? The standpoint they chose was wise: after seven generations, not only will all who made the choice be gone, but all those who heard the elaborate explanations of the choice from their grandparents will also be gone. The choice will have to stand on its own, without any defenders beyond its own wisdom. That may be the best test of "sustainability": that seven generations away, a path chosen still seems to be the correct one. That is the perspective we are looking for in this exploration.

Ethics is usually all about us. We are animals, who experience pleasure and pain, need and satisfaction; we are social, and need a rule-governed regime that satisfies our sense of justice; and we are free, rational, able to discern duties and take responsibility for our actions. From those three facts about us, we are able to see the general pattern of our intra-species obligations and the source of all the conflicts.

We know that one of the general duties we all have is the stewardship of the natural environment; but how do we know that? Our duty to non-human nature would have to be a duty of stewardship proper—there is a world that needs to be cared for, and intelligent human beings are the only ones who can do it. That's us. No simple ethic will capture our duties to Nature in a traditional theory, for the theories were all developed to treat of human beings only. "Each one to count as one and none to count as more than one," the first principle of equality and justice, dissolves in confusion if we try to extend it to chimpanzees, woodchucks, mosquitoes and bacteria. Utilitarianism, the only philosophy that truly supports rights for

---

[11] Wikipedia, "Seven Generation Sustainability." Accessed November 15, 2012.

non-human animals, asks that we seek the greatest happiness of the greatest number in the long run. Happiness of what? The costs and benefits of whaling come out very differently if you count the happiness of the whales into the equation.

It may be that the duties we have to the environment can be turned into partnership with it; nature is ingenious, and so are we, and the limitless capacity of human beings for innovation shows off to very good advantage in a system that rewards marketable innovations. The effort to integrate sound environmental philosophy with sound business strategy is not new. As early as 1988, in his pioneering *Environmental Ethics,* Holmes Rolston III argued that corporations must take into account the integrity of the natural environment, and that it was possible, contrary to the prejudices of many business executives, to incorporate a naturalistic environmental ethic in the workings of the profitable firm.[12] There are many avenues now available to business that lead to higher profits through more sensitive cooperation with the environment, and many more opening up for the future. The companies that best understand the workings of the natural environment will be the best positioned to take advantage of those opportunities.

Rachel Carson was not the first to point out that our behavior toward nature would be determined by what we saw when we looked at it; Aldo Leopold has that honor.[13] In 1968, Senegalese ecologist Baba Dioum summed it up: "In the end, we will conserve only what we love; we will love only what we understand; and we will understand only what we are taught."[14] But Carson, summing up the conclusions of *Silent Spring,* put the problems of our current approach to nature with peculiar elegance:

> The "control of nature" is a phrase conceived in arrogance, born of the Neanderthal age of biology and philosophy, when it was supposed that nature exists for the convenience of man. The concepts and practices of applied entomology for the most part date from that Stone Age of science. It is our alarming misfortune that so primitive a science has armed itself with the most modern and terrible weapons, and that in turning them against the insects it has also turned them against the earth.[15]

The way Carson's adversaries saw Nature is simplicity itself—Nature is our servant, or a big box of resources that belongs to us, to be controlled for our desires. But there are other ways of seeing Nature. Let's look at eight of them:

---

[12] Holmes Rolston III, *Environmental Ethics: Duties to and Values in the Natural World,* Philadelphia: Temple University Press, 1988. See especially Chap. 8.

[13] The references are to Aldo Leopold, A *Sand County Almanac And Sketches Here and There,* Special Commemorative Edition, New York: Oxford University Press, 1949, 1987, and Rachel Carson, *Silent Spring,* commemorative edition with an Introduction by Vice President Al Gore, Boston: Houghton Mifflin Company, 1962, 1994, but especially the book that she was working on when she died, *Help Your Child to Wonder.*

[14] Cited in Elliot Norse, "Marine Environmental Ethics," from *Values at Sea: Ethics for the Marine Environment,* ed. Dorinda G. Dallmeyer, Univ of Georgia Press, 2003. Cited in *The Environmental Ethics & Policy Book,* ed. Donald VanDeVeer and Christine Pierce, 3rd edition, Belmont, CA: Wadsworth, 2003, p. 240.

[15] Rachel Carson, *Silent Spring,* op.cit. p. 297.

Eight Ways of Seeing

We will describe a descending (or ascending, depending on your perspective) series of possible orientations to the natural environment. The following "positions," or orientations, on the natural environment are cobbled together from a very rich and varied literature, and there is a certain arbitrariness in any such spectrum that must set the scholar's teeth on edge. But then, this book is not written for scholars (at least not for scholars in the field of environmental thought). For our very practical purposes, the classification may help us get a handle on an otherwise overwhelming body of thought. For convenience, the items on the spectrum are ranked from Least Appreciative of the natural world to Most Appreciative; the first four positions are anthropocentric (that is, only human beings can be valued or valuers), the last four positions center value on something outside humanity. At the end of this discussion we will sketch some directions for profit-oriented enterprise in a market economy. (The reader will note some changes of tone in the positions that follow, as we attempt to adopt the point of view being described to give it a fair hearing.)

I. **The first, and most widespread, of the orientations** to the natural environment is the **resource orientation**: The woods are just trees, the trees are just wood, and there is no reason for us not to cut the wood for our purposes. Two subgroups are part of this orientation:

1. **Unlimited exploitation**: on this principle, humans take what they want, and see no limits in their appropriation from the natural world absolutely anything that might be of use, materially or symbolically. This approach to Nature is currently unfashionable, but is still the rule for environmental behavior over most of the world; it has persisted since the origin of agriculture, when recognition of the possibility of mastery, dominion, over the natural world replaced "animism," the pagan recognition of utter dependence on nature. This perception seems to arise naturally in any successful society; more poignantly, it is the only orientation possible anywhere in the world where people live at the edge of starvation, and that includes, to our everlasting shame, much of the world right now. People who need firewood for their evening fire will not heed "Please do not disturb the forest" signs, nor will the hungry respect endangered species. There is nothing we can do to change such perceptions without feeding the people, for survival comes first. The survival imperative is not the only source of terrible environmental damage; in Russia and Eastern Europe during the ascendancy of the Soviet Union, heavy industry polluted land, water and air without limit. In that case, an overpowering ideology created the perception of a need to industrialize areas that had been governed only by peasant cooperatives or remote lords, with no history of responsible self-government. A mindless agenda of industrial production led to a managerial orientation of damn-the-waste-stream, full speed ahead, and left Eastern Europe with a very badly treated landscape. Throughout the history of business in the US, for that matter, most of the emphasis has been on the increase of profit or shareholder wealth, with environmental damage treated as an "externality" of no concern to the manager.

In a highly competitive international business environment, corporations might argue still today that there is just no room in their survival plan for caring for the trees or land they need for the business. Theists of varying persuasion tend to extract a human right to treat Nature this way, as no more than raw materials and a convenient dump, from the Bible, the book of *Genesis,* Chap. 1, where Adam is given by God a right of dominion over all non-human nature.

2. **Wise Use**. The name of this orientation, the original conservation movement, was chosen by its first exponent, Gifford Pinchot, President Theodore Roosevelt's chief of forestry and founder of the Yale School of Forestry.[16] Resources are there to be used, on this orientation, and the natural environment is to be seen as nothing but resources for our use, but they must not be wasted. We must provide not only for our own generation but for our children, too. Like any prudent householder, we must conserve natural resources for future use, and make the most efficient use of those we have. So Pinchot had forests set aside, to be managed in order to supply wood for construction materials and other products into the indefinite future. While the woods were standing waiting to be cut, they should be managed for other purposes—watershed protection and recreational uses, for instance. Thus the multi-purpose National Forests were born. Pinchot argued for nothing more or less than enlightened self interest, where "interest" was defined purely economically. If we are enlightened about keeping pollution under control, for instance, we should be able to have clean air and water without damaging our lifestyle. There had been a predecessor movement, on which Pinchot drew, beginning in the 1870s: as industrialization and settlement advanced west, wealthy families of who enjoyed the chase saw their hunting grounds begin to disappear, so purchased and preserved large tracts of woods for hunting, to be held by their clubs or associations. When Roosevelt and Pinchot first started talking about setting land aside, then, to keep for the future, a significant number of the people who mattered, including Roosevelt's family, already approved of the concept.

II. **The second orientation**, one that the US only discovered in the past century and struggles with still, is a **protection orientation**, one that seeks to conserve wildness wherever it is found. Again, there are two branches:

3. **Conservation**: Mark Sagoff is a powerful spokesman for an orientation of aesthetic or spiritual conservation (as opposed to Pinchot's economic conservation). Nature is perceived as threatened and as valuable, requiring our protection. We should be willing to modify our extravagant lifestyles in order to save scarce valuable resources. First of all, the hunting of endangered large animals for commercial purposes should be stopped (elephants for their ivory, tigers for their pelts, whales and dolphins for cat food). The preservation of these beautiful animals is much more important than the indulgence of tastes for exotic rugs and carvings. Second, there should be parks, where future generations can enjoy unspoiled

---

[16] The name is problematic, because recently (in the last years of the twentieth century) an anti-environmental movement of the same name has emerged, fueled by the philosophy of Milton Friedman and the subsidies of the less enlightened extractive industries. There is no relation between Gifford Pinchot's philosophy and the new organization.

nature; parks should be the appropriate monuments of our nation, corresponding to the great castles and cathedrals of Europe, available to the American people and the world in perpetuity for rest, recreation, education and spiritual inspiration. Further, there should be quiet but firm pressure on all consumers to buy less polluting cars, to install energy-saving windows, turn out lights and turn off computers, recycle all materials that can be recycled, and so forth. Our most valuable resources should be available not only for private profit, but for public use. The natural world has limits, which we are approaching, and we must work to eliminate the wasteful practices that damage it. We must remember that we are responsible for keeping the world beautiful and healthy for our grandchildren, and support organizations that engage in responsible educational activities to teach young people the value of nature and the ways of preserving it. Theists of this persuasion also derive their authorization from the book of *Genesis,* but from Chap. 2, where God puts humans in the Garden "to tend and to keep it," imposing upon us a duty of stewardship.

4. **Preservation**: This approach, traced to John Muir (the dedicated mountaineer who spent much of his career trying to prevent the damming of Hetch Hetchy), attributes to natural tracts a life and value of their own, and requires us to preserve them for that value. Proponents insist that the Conservation orientation, economic or otherwise, will never succeed in protecting the natural world. As soon as it is suggested that wilderness should be "accessible to the public," we have created a dilemma between wilderness preservation and public access, and preservation is unlikely to win. Already, only a century or so after their founding, our National Parks are being "loved to death" by a public all too eager to "appreciate nature" and all too unwilling to leave their preferred lifestyle behind. Asphalt roads are required, enormous parking lots with electrical fixtures and plumbing to accommodate the recreational vehicles assembling in flotillas every summer, trash cans, rest rooms and concession stands for food and souvenirs. Handicapped ramps are required, and places where wheelchairs may be brought to carved out "scenic points," carefully fenced, so that their users may enjoy the view. The wilderness that spanned the country when our forefathers arrived is not being preserved by such "parks." Future generations will be best served by inheriting intact ecosystems, large expanses of forest that will remain untouched forever, available to people who want to walk in, pack their food in (and pack their trash out), and just enjoy the woods as they always have been and (as long as we don't cut them) always will be. They are not just for recreation; these ecosystems are our future universities on the workings of the natural world, and our storehouses for generations of pharmaceuticals, foods and other materials not yet discovered. We cannot afford to lose them. No roads are necessary, nor any special care; the forest can tend itself just as it always has (including surviving forest fires). When we have conducted polls on the subject, a substantial number of people have answered that they are in favor of the preservation of Wilderness Areas not for any projected use of theirs but just to know that they are there, forever, for their grandchildren's grandchildren and on to the seventh generation. The preservation of untouched nature has what can only be called a moral value, a satisfaction built into the stewardship itself.

III. **The third orientation** is a **rights orientation**, growing out of our increasing national habit of settling whatever conflicts we have by going to court and arguing them out as entitlements. Here, we are talking about moral rights as well as legal rights. What, beyond human beings, might be "morally considerable," or "entitled" to respect? There are two answers:

5. Attribution of **rights to animals**: The "animal rights" orientation has several branches and is difficult to summarize; let us attempt. The most familiar, associated with Peter Singer, stems from Utilitarianism: as (the higher) animals can suffer, they should not, as a matter of right, be subjected to painful or confining circumstances for human purposes. It is not sufficient to hold the simple "humane treatment" orientation, that we conduct ourselves as virtuous human beings when we are kind to our animals, who appreciate our kindness but have no right to it. Animal Rights advocates point out that southern slaveholders were urged to be kind to their slaves, too, but kindness is just not the same thing as protection of one's rights under law. If we take animal welfare seriously at all, we must go beyond the patchwork "cruelty to animals" statutes to a robust concept of animal rights. They do not urge that non-human creatures be given the full panoply of civil rights, from non-discrimination to the franchise; animals have essentially the rights of children, not to participate in the political community, but to be protected from harm by its members. Tom Regan, on the other hand, holds that all creatures who can be "subjects of a life," subjects of an existence that can go better or worse for them, deserve to have their interests in living a better life respected. Whatever their theoretical stance, animal rights advocates agree on certain practical conclusions. First, save in circumstances where it would be equally acceptable to use babies as subjects, animals should not be used in scientific research. Second, animals raised for food should live in freedom, in natural settings, drug free and eating food that would be theirs in nature, prior to slaughter. Third, pets and working animals (racehorses, hunting or racing dogs, participants in animal acts) should be treated well, and cared for or humanely put down when they are no longer useful in their work. Fourth, many advocates for animal rights (but not all) would argue that animals with special mental capacities (whales, dolphins, and the Great Apes, primarily) that remind us eerily of ourselves—the most intelligent animals—deserve special rights of preservation and care; not only should we care about those temporarily in our power (dolphins in Sea World, for instance, or chimps in traveling shows), but to the greatest possible extent, we should protect the habitat of these animals so that they can live together in freedom according to their own laws. *Many of the animal rights persuasion, but by no means all, think that we should all be vegetarians*; some ("vegans") believe that we should not eat, or use in any way, any animal products at all.

6. Attribution of **rights to ecosystems**: Should Trees Have Standing? Christopher Stone once asked, in response to a case where the trees were threatened but it was difficult to find a threatened human.[17] In *Sierra v. Morton*, the case

---

[17] Christopher Stone, *Should Trees Have Standing?*, Oxford University Press, third edition 2010. Originally published 1972, became a rallying point for the environmental movement.

in question and a classic of environmental law, the Sierra Club sued to prevent Disney Enterprises from developing an amusement park and ski area on Mineral King, a particularly beautiful stretch of California mountains. They argued that the mountain itself and the wild area that depended on it had a right not to be destroyed, and they claimed only to represent it. The judges threw out the case, and required the Sierra Club to come back into court with some human beings who would be hurt by the development (which it did, and won). But the claim retains its interest. The fundamental living unit of the natural environment is not the individual organism, but the ecosystem, the entire system of biological relationships that make it possible for the organisms in it to live and thrive. Therefore, if we wish to preserve environmental values, we should be concerned to preserve whole ecosystems (forests, deserts, and wetlands, for example). We have learned, as per the argument above in the Animal Rights heading, that protection of anything, human or otherwise, entails assigning legal rights to that thing, and the right to sue, itself or through a proxy, for its own entitlements. To this end, ecosystems should be permitted to sue for protection in their own names, through the advocacy of environmental associations.

IV. **The fourth orientation** is **holistic**, starting with the notion that nature as a whole is a complete and privileged system or organism, and that we as humans are bound by obligations of contract or participation to defer to its interests.

7. **Community holism**: This category contains a variety of perspectives, having in common the land ethic of Aldo Leopold, which "changes the role of Homo sapiens from conqueror of the land-community to plain member and citizen of it."[18] Egalitarian relationships with plants and animals may take the form of mutual accommodation over centuries—between, for instance, shepherds and their flocks, farmers and their oxen, all of us and our dogs. The fundamental insight of this orientation is that we and the natural world evolved together, and support each other by mutual forbearance and cooperation, acknowledging each other's right to exist and to flourish. It is not surprising that this mutual recognition can turn into love, affection; the Pygmies of Central Africa insist that the forest gives them affection as well as providing them with food. E. O. Wilson and Stephen Kellert have pioneered studies in "biophilia," the dependence phenomenon of human health and wholeness with association with the natural world.[19] Community holism begins with an appreciation of the land, which entails seeing it as the organic whole that it is. It is not a blank slate on which we can write at will, but a living, breathing community with its own laws and understandings, which we must learn. The unseeing, the ignorant, regard the land as simply ground to be worked (or trampled), and regard holism as so much fanciful anthropomorphism, but they are wrong. There is important knowledge to be gained from Nature, if we will but be patient, look, and learn to interpret what we see.

---

[18] Aldo Leopold, op.cit, p. 204.

[19] Edward O. Wilson, *Biophilia,* Cambridge: Harvard University Press, 1984.

8. **Deep Ecology**: This view, initiated by some very original thinkers (Arne Naess, Bill Devall and George Sessions, among others), is difficult to pin down. Its simplest interpretation is that the living organisms of the world are so many organs in the single organism of the biosphere, and that we must therefore live subject to the organic laws of that system. (In some streams of radical environmentalism, but by no means all, this biosphere is designated "Gaia," the earth-goddess.) We are the brain in this organism. We have the ability to discern the workings of the whole, to see what we must do to preserve its health and beauty, and to behave accordingly. Our moral development is found in increasing identification with all living things.[20] In theory, should severe injury occur to the earth (say, from contact with an asteroid), we would have the responsibility to repair the damage, if the earth could not do it itself. Above all, we have a duty, if we can call it that, to discover the laws of the biosphere as they apply to us and abide by them, not interfering with the other members of the biosphere or attempting to modify or thwart any of its purposes.

Of all these orientations, the "rights" perspectives are the first to make major claims on our attention, and are already affecting our legal system. Why does it make sense to attribute rights to animals? In the philosophical literature, most discussions of this question turn on some "feature" of animals that is supposed to link them with us, or more likely, separate them from us. Syllogisms abound, of the sort, "All creatures that can suffer (like humans) are worthy of moral consideration; the higher animals can suffer; therefore the higher animals are worthy of moral consideration." (Jeremy Bentham reasoned so, putting the matter much more eloquently.) Or more likely, "Only creatures that can reason are worthy of moral consideration; non-human animals cannot reason; therefore non-human animals are not worthy of moral consideration." (Those syllogisms are not strictly in form, but they're valid.) Such reasoning is easy to tear apart. There turns out to be no simple way to separate the "higher animals" (those with nervous systems like ours) from the "lower animals," and if we have pledged ourselves to cause no suffering, where do we stop? Scallops can't get arthritis, but can they feel pain when they are scooped out of their shells? If, on the other hand, we restrict moral consideration to creatures that can reason, we exclude infants, the demented elderly, the emotionally disturbed (for the most part), and a wide range of developmentally disabled humans. The rights claimed for animals are precisely those extended to such humans—to be taken care of, fed and sheltered if they are in our power, and treated respectfully. Incidentally, the history of the philosophical treatment of this question is not encouraging; there simply is no "set of characteristics" that distinguish the human from the non-human. Are animals valuable? There are other examples of simple difference in perspective on the subject of the natural environment, but none so clear and so potentially divisive. Let's look at that potentiality.

---

[20] Guidance in this section especially, and throughout the chapter, was provided by John Nolt of the University of Tennessee at Knoxville.

Animal rights advocates are not just intellectually convinced that whatever it is that makes humans worthwhile also makes animals worthwhile. Many of them genuinely love and respect the animals they defend, so they are genuinely passionate in their defense of their perspective. They tend to regard those who disagree with them as "speciesists," on the analogy of racists, willfully blind to the natural worth of animals, blinding themselves for the sake of continuing the exploitation, to serve their own convenience and pleasure. Animal rights advocates perceive a being that has value, moral considerability equal to our own, where others do not. In this respect, their advocacy is much like the anti-abortion faction of the American public now, who see a being of equal value in the unborn human where others do not, and they have been compared to abolitionists on the slavery question. As in both those similar campaigns, the potential for violence is not far from the surface, and several Animal Rights factions (including the Animal Liberation Front, ALF, and PETA, People for the Equal Treatment of Animals) have engaged in criminal trespass and other violations of law to "liberate" animals from laboratories and farms. We find much the same potential for the defense of ecosystems. Violence is not as likely in defense of trees, which lack soft fur and appealing brown eyes, but in the redwood forests of the west coast, where confrontation has been the rule for too many years, tree-sitters and Earth First! Saboteurs, in their "monkey-wrenching," have courted criminal proceedings as a regular tactic.

The rights orientation asks primarily for protective laws, and where animals are concerned, they have got laws on the books. On the simple hypothesis that animals have a right to be cared for humanely, all animals used in laboratory experiments are now subject to strict regulations regarding veterinary care, cage space, food and "cruelty"—any infliction of pain beyond the needs of the research being done, any signs of filth or neglect, any evidence (from tapes of the experiments, for instance) of callousness or enjoyment of the animal's suffering. Prohibitions on cruelty to animals dating back to the middle of the nineteenth century forbid the public abuse of domestic animals (whipping, for instance), and serious neglect of animals in our care (starvation, filthy quarters, general poor health). No statute forbids us to kill any animal in our care, as long as it is done humanely, but laboratory and domestic animals at least have laws to hold their owners to account for treatment. (Campaigns on behalf of farm animals have not enjoyed the same success.) Neither in the case of animals nor in the case of ecosystems do we have any law in place that attributes rights to the natural object to sue in its own behalf—neither animals nor ecosystems have legal standing—but laws restricting human activity that may harm natural objects are very much with us.

Let us note, and keep in mind, the conceptual gulf between perspectives 3 and 4, conservation and preservation. Conservation has us protect the environment only for human use, but along dimensions that include many human values beyond the purely economic—recreational, aesthetic, mental and spiritual health. This ambivalence lands it in a perpetual dilemma between preserving the environment and ensuring public appreciation, with all the parking lots, rest stops, RV hookups and internet cafés entailed by that responsibility. Preservation assigns value, not to the ecosystem being preserved in itself (that is the goal of the ecosystem rights move, perspective 6), but to the value *we* place on having that ecosystem preserved, as it is, for our grandchildren's grandchildren, and to the seventh generation. Yet both these orientations form part of our approach to the environment, and both are valid.

The Case for Environmentalism

Environmentalism is not just a political position, "cause," or "special interest." By definition, it cannot be a special interest, like the sugar lobby or the hotel workers' union—the economic interests of the environmentalists are not at stake. Along with other contemporary causes adopted simply because they are good causes—feeding the poor of the world (Oxfam), or the protection of human rights (Amnesty International), for instance—environmentalism conceptually excludes the link between advocacy and economic advantage typical of the special interest group. (The homeowner who wants the woods bordering his property protected in order to preserve his property value is *ipso facto* not an environmentalist, at least not when he advocates for "protection" for those woods.) There is no reason why environmental measures cannot also create economic advantages. But for present purposes we are exploring the non-economic side of an increasingly popular movement, with coherent purposes and undeniable influence on the global scene.

Nor is environmentalism just a political position, adopted from a vision of the Public Good. It is that, but for its participants it can be considerably more: it can be a life-absorbing care and dedication, like a family, and it can be, or fill the place of, a religion, deserving devotion and service even when external indicators suggest the cause is hopeless. Given the emotional power of the commitment, it's worth an attempt to understand just where the movement is coming from.

How does the environmentalist see the present situation? Like this:

We are all in a car, hurtling down the mountain toward the cliffs from which the car will plunge into the bottomless ocean. We are all passengers in the rear seat, while the front seat holds the corporations and the governments, chatting amiably. It's hard to see who's driving, but the likely end of the trip is very clear. Please, sirs, could we stop? Or at least, slow way down? *Now*, please?

Somewhat surprisingly, this perspective is not shared by the folks in front. They seem to think that the car is going *up* the mountain, in a course of human progress charted by their great mapmaker Adam Smith, who believed that cutting down forests and clearing fields and damming rivers all made nations wealthier. They don't. They destroy the working capital of the nation and mortgage its future. But the folks in the front either do not know this, or they do not care. They are doing very well on this route—which they see as Creating Value and the environmentalist sees as Planned Plunder—and they see no need to stop the car. They think the complaining passengers are just a bit nutty, corporation-haters, tree-huggers, political ideologues, or worse.

To be sure, most of the passengers in the rear are not complaining. The vast majority of them are so concerned with day-to-day survival that they have no time to complain about anything, and a solid number of those who remain are convinced (probably by paid agents of the folks in front) that their livelihoods depend on opposing the nuts that are complaining. But the complainers see with crystal clarity the decline of the conditions for human existence and indeed all life on the planet, and they are not to be deterred; they have the numbers on their side. They see the situation for humans and the planet as somewhere between very serious and desperate, they see their role as saving a world from a small group of insensitive and selfish schemers, malefactors of great wealth, and they are willing to put into their work the energy appropriate to such a battle. People who are convinced of the rightness of their cause are unlikely to flinch from dying in battle. It might be worth remembering that.

The problem is not that the Environmentalist has a "political position" that he refuses to "compromise." True, the genius of democratic governments is an understanding of compromise (consider the history of labor negotiations). But compromise is unavailable, given the nature of environmental problems. Recall the argument above: Consider the logging of an old-growth forest. If the environmentalist reluctantly agrees to let the timber folks take 50 % of the forests, on condition that they leave the rest, there goes 50 % of the forest. But loggers don't just need jobs this year. They need jobs every year. Two years from now, when the 50 % they took is all gone, they're going to come back for 50 % of the rest. And the year after, when that's gone, they will ask for another "compromise." Old growth redwood forests, at least, do not regrow, even if (as is not always the case) they are replanted, not for 1000 years. Nor do toxic wastes, like PCBs or pesticides, go away; they are bioaccumulative, persist and get worse, in the body of the human and in the body of the earth. "Compromise" compromises the future of the earth and the human race with it; the environmentalist cannot compromise.

What does the environmentalist want? He wants the salvation of the world. The car is hurtling toward its destruction, and he wants to stop it. If he cannot do that, he at least wants to slow it down, in hopes that the time purchased by slowing it may provide the opportunity for the drivers, or at least a larger number of the passengers, to realize that they are on the wrong road. All the clanking machinery of government—the Environmental Protection Agency (EPA), the state and local Departments of Environmental Protection and Conservation Departments, the Clean Air Act, the Clean Water Act, Superfund (CERCLA and SARA), the recycling requirements on federal, state and local levels—are stopgap measures. As William McDonough points out somewhere, when you're trying to go from Washington to Baltimore, and you find yourself on the way to Atlanta, slowing from 90 miles an hour to 45 miles an hour won't really solve your problem; you need to turn around and go the right direction. But if the driver continues to refuse to look at a road map or ask directions (men!), at least slowing down will mean that when you do finally turn around, you won't have so far to go. And if you lose the battle, and they never do turn the car around, at least your grandchildren will live slightly longer and happier lives than they would have if that first speed had been maintained.

What can be expected of the environmentalists and the numerous organizations that they support? They cannot be expected to go away, or come round to the driver's view that the earth's carrying capacity will somehow expand to allow a population of 9 billion (foreseen for the middle of this century) to live at the standard of the suburban North. They can be expected to work for future environmental protection legislation; they can be expected to constitute a disconcerting global presence, influencing international dealings to an extent not predicted by their U.S. influence; and they can be expected to work through Civil Society Organizations (CSOs; also known as private voluntary organizations or non-governmental organizations, NGOs) like Greenpeace, which have attained an outsized world influence through coordinated and dramatic environmental campaigns.

The problem right now is to achieve some sort of engagement in the dialogue with the people in the front seat. There are two approaches to dialogue. One is confrontative (shoot out the tires). The preferred approach is cooperative. For that we need to refocus the perspective as an opportunity to get something good done. For the near and middle future, hope lies in the increasing realization that the natural environment must be preserved as a condition for the continuation of any successful enterprise, business included.

WITHDRAWN